Studies in Fuzziness and Soft Computing

Volume 378

Series editor

Janusz Kacprzyk, Polish Academy of Sciences, Systems Research Institute,
Warsaw, Poland
e-mail: kacprzyk@ibspan.waw.pl

The series "Studies in Fuzziness and Soft Computing" contains publications on various topics in the area of soft computing, which include fuzzy sets, rough sets, neural networks, evolutionary computation, probabilistic and evidential reasoning, multi-valued logic, and related fields. The publications within "Studies in Fuzziness and Soft Computing" are primarily monographs and edited volumes. They cover significant recent developments in the field, both of a foundational and applicable character. An important feature of the series is its short publication time and world-wide distribution. This permits a rapid and broad dissemination of research results.

Indexed by ISI, DBLP and Ulrichs, SCOPUS, Zentralblatt Math, GeoRef, Current Mathematical Publications, IngentaConnect, MetaPress and Springerlink. The books of the series are submitted for indexing to Web of Science.

More information about this series at http://www.springer.com/series/2941

Urszula Bentkowska

Interval-Valued Methods in Classifications and Decisions

Springer

Urszula Bentkowska
Faculty of Mathematics and Natural
Sciences
University of Rzeszów
Rzeszów, Poland

ISSN 1434-9922 ISSN 1860-0808 (electronic)
Studies in Fuzziness and Soft Computing
ISBN 978-3-030-12929-3 ISBN 978-3-030-12927-9 (eBook)
https://doi.org/10.1007/978-3-030-12927-9

Library of Congress Control Number: 2019930980

This Springer imprint is published by the registered company Springer Nature Switzerland AG
The registered company address is: Gewerbestrasse 11, 6330 Cham, Switzerland

To my Family and Friends,
especially to my Parents

Preface

> *I saw the need, but I did not know how to satisfy it. I posed the problem to my best friends, Herbert Robins and Richard Bellman, because as mathematicians they were better qualified than I was to come up with a theory which was needed. Both were too busy with their own problems. I was left on my own.*
>
> Lotfi A. Zadeh [1]

Since the seminal paper on fuzzy sets was published [2], plenty of books and papers devoted to the topic of fuzzy sets theory, its extensions and applications appeared. According to the Web of Science, there are over 198,000 works with the *fuzzy* as a topic. Among them, there are also works of the best friends mentioned in the quotation by Lotfi Zadeh (e.g., [3]). The author of this monograph also would like to contribute to the subject of fuzzy sets, especially interval-valued fuzzy sets, which are one of the most important and developing generalizations of the fuzzy sets theory. However, the presented results may be also advantageous to the whole community, not only fuzzy, but more generally involved in research under uncertainty or imperfect information.

Fuzzy sets theory and its extensions are interesting not only from a theoretical point of view, but also they have applications in many disciplines such as computer science and technology. Fuzzy sets turned out to be effective tools for many practical applications in all areas, where we deal with natural language and perceptions. Fuzzy sets and fuzzy logic contributed to the development of the artificial intelligence and its applications. Fuzzy sets theory and its diverse extensions are still one of the most important approaches for dealing with uncertain, incomplete, imprecise, or vague information. The aspect of data uncertainty is studied intensively in many contexts and scientific disciplines. Many different forms of uncertainty in data have been recognized. Some come from conflicting or incomplete information, as well as from multiple interpretations of some phenomenon. Other arise from lack of well-defined distinctions or from imprecise boundaries. It is impossible to eliminate completely uncertainty and ignorance from everyday experience of scientists, specialists in various fields, and also the life of an average man. According to Lotfi A. Zadeh *As complexity rises, precise statements lose meaning and meaningful statements lose precision*. This is why there is a need to develop effective algorithms and decision support systems that would be able to capture the arising problems.

The main aim of this monograph is to consider interval-valued fuzzy methods that improve the classification results and decision processes under incomplete or imprecise information. The presented results may be useful not only for the community working on fuzzy sets and their extensions, but also for researches and practitioners dealing with the problems of uncertain or imperfect information. The key part of the monograph is the description of the original classification algorithms based on interval-valued fuzzy methods. The described algorithms may be applied in decision support systems, for example, in medicine or other disciplines where the incomplete or imprecise information may appear (cf. Chap. 4), or for data sets with a very large number of objects or attributes (cf. Chap. 5). The presented solutions may cope with the challenges arising from the growth of data and information in our society since they enter the field of large-scale computing. As a result, they may enable efficient data processing. The presented applications are based on theoretical results connected with the family of comparability relations defined for intervals and other related notions. We show the origin, interpretation, and properties of the considered concepts deriving from the epistemic interpretation of intervals. Namely, the epistemic uncertainty represents the idea of partial or incomplete information. It may be described by means of a set of possible values of some quantity of interest, one of which is the right one [4]. Since the subject is wide, we mainly concentrate on theory and applications of new concepts of aggregation functions in interval-valued fuzzy settings. The theory of aggregation functions became an established area of research in the past 30 years [5]. Apart from theoretical results, there are many applications in decision sciences, artificial intelligence, fuzzy systems, or image processing. One of the challenges is to propose implementable aggregation methods (cf. [6]) to improve the usability of the proposed ideas. Such methods provide a heuristic which may be conveniently implemented and easily understood by practitioners. Moreover, another challenge is related to the ability of including in the proposed solutions human-specific features like intuition, sentiment, judgment, affect, etc. These features are expressed in natural language which is the only fully natural means of articulation and communication of the human beings. This idea led to considering aggregations inspired by the Zadeh idea *computing with words* [7]. Computing with words (CWW) (cf. [8]) has a very high application potential by its remarkable ability to represent and handle all kinds of descriptions of values, relations, handling imprecision. There are many aggregation methods that try, with success, to resolve the challenges of nowadays problems (cf. [9–16]). In this book, we examine the so-called possible and necessary aggregation functions defined for interval-valued fuzzy settings. One of the reasons to consider these types of aggregation operators is connected with the fact that these notions of aggregation functions were recently introduced [17] and they have not been widely examined before.

The book consists of two parts. In the first part, theoretical background is presented and next in the second part application results are analyzed. In theoretical part, in Chap. 1 elements of fuzzy sets theory and its extensions are provided. There are presented the notions of interval-valued fuzzy calculus. Diverse orders applicable for interval-valued comparing, including interval-valued fuzzy settings, are discussed. Furthermore, in Chap. 2 aggregation functions defined on the unit interval $[0, 1]$ are recalled and useful notions and properties are provided. Construction methods of interval-valued aggregation functions derive from the real-line settings and interval-valued aggregation functions often inherit the properties of their component functions defined on the unit interval $[0, 1]$. All these issues will be presented in Chap. 2.

Part II covers two major topics: decision-making and classification problems. Chapter 3 is devoted to decision-making problems with interval-valued fuzzy methods involved. It is pointed out the usage of new concepts with possible and necessary interpretation involved. Next, the classification problems are discussed. When classifiers are used there is a problem of lowering its performance due to the large number of objects or attributes and in the case of missing values in attribute data. In this book, it is shown that in such situations interval-valued fuzzy methods help to retrieve the information and to improve the quality of classification. These issues are discussed in Chaps. 4 and 5. In Chap. 4, there are proposed methods of optimization problem of *k-NN* classifiers that may be useful in diverse computer support systems facing the problem of missing values in data sets. Missing values appear very often in data sets of computer support systems designed for the medical diagnosis, where the lack of data may be due to financial reasons or the lack of a specific medical equipment in a given medical center. Chapter 5 presents methods of dealing with large-scale problems such as large number of objects or attributes in data sets. Specifically, there is presented a method of optimization problem of *k-NN* classifiers in DNA microarray methods for identification of marker genes, where typically there is faced the problem of huge number of attributes. Finally, in Chap. 6, there is presented the performance of the new types of aggregation functions for interval-valued fuzzy settings in the computer support system OvaExpert [18]. The book ends with a brief description of the future research plans in the area of presented problems, both in the theoretical and practical aspects.

The book is aimed at practitioners working in the areas of classification and decision-making under uncertainty, especially in medical diagnosis. It can serve as a brief introduction into the theory of aggregation functions for interval-valued fuzzy settings and application in decision-making and classification problems. It can also be used as supplementary reading for the students of mathematics and computer science. Moreover, the results on aggregation functions may be interesting for computer scientists, system architects, knowledge engineers, programmers, who face a problem of combining various inputs into a single output. The classification algorithms considered in this book (in Chaps. 4 and 5), along with other supplementary materials are available at [19], where there are provided suitable files to download and run the experiments.

I would like to thank Prof. Józef Drewniak for introducing me to the subject of fuzzy sets theory. Moreover, I would like to thank other Professors that helped me in better understanding the nuances of fuzzy sets theory, its extensions, and applications. Namely, these are the following persons (listed in the alphabetical order): Jan G. Bazan, Humberto Bustince, Bernard De Baets, Przemysław Grzegorzewski, Janusz Kacprzyk, Radko Mesiar, Vilém Novák, and Eulalia Szmidt. I am also grateful to my colleagues from Poland and abroad with whom I cooperated working on scientific problems or whom I met during scientific conferences. Especially, I would like to thank my colleagues from the University of Rzeszów with whom we spent many hours on seminars discussing scientific problems.

Finally, I would like to express my deepest gratitude to my family and friends for their constant encouragement and support.

Rzeszów, Poland
October 2018

Urszula Bentkowska

References

1. Zadeh, L.A.: Fuzzy logic–a personal perspective. Fuzzy Sets Syst. **281**, 4–20 (2015)
2. Zadeh, L.A.: Fuzzy sets. Inf. Control **8**, 338–353 (1965)
3. Bellman, R., Giertz, M.: On the analytic formalism of the theory of fuzzy sets. Inf. Sci. **5**, 149–156 (1973)
4. Dubois, D., Prade, H.: Gradualness, uncertainty and bipolarity: making sense of fuzzy sets. Fuzzy Sets Syst. **192**, 3–24 (2012)
5. Beliakov, G., Bustince, H., Calvo, T.: A Practical Guide to Averaging Functions. Studies in Fuzziness and Soft Computing. Springer International Publishing, Switzerland (2016)
6. Albers, S.: Optimizable and implementable aggregate response modeling for marketing decision support. Int. J. Res. Mark. **29**, 111–122 (2012)
7. Kacprzyk, J., Merigó, J.M., Yager, R.R.: Aggregation and linguistic data summaries: a new perspective on inspirations from Zadeh's fuzzy logic and computing with words. IEEE Computational Intelligence Magazine (forthcoming) (2018)
8. Zadeh L.A.: Computing with Words—Principal Concepts and Ideas. Studies in Fuzziness and Soft Computing, p. 277, Springer (2012)
9. Blanco-Mesa, F., Merigó, J.M., Kacprzyk, J.: Bonferroni means with distance measures and the adequacy coefficient in entrepreneurial group theory. Knowl.-Based Syst. **111**, 217–227 (2016)
10. Castro, E.L., Ochoa, E.A, Merigó, J.M., Gil Lafuente, A.M.: Heavy moving averages and their application in econometric forecasting. Cybern. Syst. **49**(1), 26–43 (2018)
11. Castro, E.L., Ochoa, E.A, Merigó, J.M.: Induced heavy moving averages. Int. J. Intell. Syst. **33**(9), 1823–1839 (2018)
12. Liu, P., Liu, J., Merigó, J.M.: Partitioned Heronian means based on linguistic intuitionistic fuzzy numbers for dealing with multi-attribute group decision making. Appl. Soft Comput. **62**, 395–422 (2018)
13. Merigó, J.M., Palacios Marqu´es, D., Soto-Acosta, P.: Distance measures, weighted averages, OWA operators and Bonferroni means. Appl. Soft Comput. **50**, 356–366 (2017)

14. Merigó, J.M., Gil Lafuente, A.M., Yu, D., Llopis-Albert, C.: Fuzzy decision making in complex frameworks with generalized aggregation operators. Appl. Soft Comput. **68**, 314–321 (2018)
15. Merigó, J.M., Zhou, L., Yu, D., Alrajeh, N., Alnowibet, K.: Probabilistic OWA distances applied to asset management. Soft Comput. **22**(15), 4855–4878 (2018)
16. Zeng, S., Merigó, J.M., Palacios Marqu´es, D., Jin, H., Gu, F.: Intuitionistic fuzzy induce-dordered weighted averaging distance operator and its application to decision making. J. Intell. Fuzzy Syst. **32**(1), 11–22 (2017)
17. Bentkowska, U.: New types of aggregation functions for interval-valued fuzzy setting and preservation of pos-B and nec-B-transitivity in decision making problems. Inf. Sci. **424**, 385–399 (2018)
18. Dyczkowski, K.: Intelligent Medical Decision Support System Based on Imperfect Information. The Case of Ovarian Tumor Diagnosis. Studies in Computational Intelligence, Springer (2018)
19. http://diagres.ur.edu.pl/ ~ fuzzydataminer/

Contents

Part I
Foundations

In this part there are presented concepts related to fuzzy sets theory and its extensions. Moreover, short historical mentions regarding the development of these notions are provided. There are recalled diverse types of comparability relations between intervals, including orders and linear orders. Mostly, this part of book concerns aggregation functions defined both on the real line (or the unit interval [0,1]) and for interval-valued settings. There are presented diverse representation methods of interval-valued aggregation operators, their properties, construction methods and dependencies between diverse classes of these operators. The considered aggregation operators may fulfil various monotonicity conditions, namely with respect to the classical partial order, with respect to the linear orders or with respect to the two other distinguished comparability relations derived from the epistemic interpretation of intervals. Special attention is paid to the recently introduced notions of pos-aggregation functions and nec-aggregation functions. There is also discussed the problem of preservation of width of intervals by aggregation operators.

Chapter 1
Fuzzy Sets and Their Extensions

In July of 1964, I was in New York, and was scheduled to have dinner with my friends. The dinner was canceled. I had a free evening. My thoughts turned to the issue of cointensive indefinability. At that point, a simple idea clicked in my mind the concept of a grade of membership. The concept of a grade of membership was a key to the development of the theory of fuzzy sets.

Lotfi A. Zadeh [1]

In this chapter basic notions regarding fuzzy calculus, its history and basic properties are recalled. Moreover, extensions of fuzzy sets are briefly described and the most important results concerning interval-valued fuzzy calculus are provided. Especially, the notions of diverse order and comparability relations for interval-valued settings are discussed.

1.1 Elements of Fuzzy Sets Theory

Lotfi A. Zadeh is credited with inventing the specific idea of a *fuzzy set*, i.e. an extension of the classical notion of set in his seminal paper on fuzzy sets [2]. He gave a formal mathematical representation of the concept which was widely accepted by scholars. In fact, the German researcher Dieter Klaua independently published a German-language paper presenting examples of fuzzy sets in the same year, but he used a different terminology (he referred to *many-valued sets*, *mehrwertiger mengenlehre* in German, [3]). The ideas of a fuzzy set, as a generalized characteristic function, appeared much earlier. Jan Łukasiewicz is a pioneer with the concept of

U. Bentkowska, *Interval-Valued Methods in Classifications and Decisions*,
Studies in Fuzziness and Soft Computing 378,
https://doi.org/10.1007/978-3-030-12927-9_1

multivalued logic. In 1920 he introduced the concept of a *trivalent logic* [4]. Edward Szpilrajn considered a generalized characteristic function with the values in the Cantor Set [5]. Examples of fuzzy relations were provided by Karl Menger in 1951 (*ensembles flous* in French, [6]). Helena Rasiowa in 1964 considered generalized characteristic functions (functions $f : X \to L$, where L is any logical algebra, cf. [7]), but did not use them in the sense of fuzzy sets.

Farther important papers on fuzzy sets published after 1965 are the ones on fuzzy relations by Zadeh [8] and L-fuzzy sets by Goguen [9]. Now there exist a long list of papers on fuzzy sets theory and there exist a long list of applications such as digital cameras, fraud detection systems, fuzzy logic blood pressure monitors, fuzzy logic based train operation systems and many others. Fuzzy sets theory is a very wide branch of science. In this section we will provide only some basic notions connected with the theory of fuzzy sets.

1.1.1 Basic Notions of Fuzzy Calculus

A fuzzy set may be treated as a generalization of the notion of a characteristic function of a set $A \subset X$, namely

$$\chi_A(x) = \begin{cases} 1, & x \in A \\ 0, & x \notin A \end{cases}, \quad x \in X.$$

Definition 1.1 ([2]) A fuzzy set A on $X \neq \emptyset$ is defined as an arbitrary function $A : X \to [0, 1]$. The family of all fuzzy sets on X is denoted by $\mathcal{FS}(X)$. A value $A(x)$ is called a membership value of $x \in X$ to the fuzzy set A.

Operations on fuzzy sets are defined analogously to the operations on crisp sets.

Definition 1.2 (*cf.* [2]) A complement of a fuzzy set $A \in \mathcal{FS}(X)$ is called $A' \in \mathcal{FS}(X)$, where

$$A'(x) = 1 - A(x), \quad x \in X. \tag{1.1}$$

Inclusion, sum and intersection of $A, B \in \mathcal{FS}(X)$ are defined in the following way:

$$A \leqslant B \Leftrightarrow \forall_{x \in X} A(x) \leqslant B(x), \tag{1.2}$$

$$(A \vee B)(x) = \max(A(x), B(x)), \quad (A \wedge B)(x) = \min(A(x), B(x)), \ x \in X. \tag{1.3}$$

Special cases of fuzzy sets are fuzzy relations.

Definition 1.3 ([2]) A fuzzy relation between $X, Y \neq \emptyset$ is called an arbitrary function $R : X \times Y \to [0, 1]$. The family of all fuzzy relations between X and Y is denoted by $\mathcal{FR}(X, Y)$. If $Y = X$, then we say that R is a fuzzy relation on X. The family of all fuzzy relations on X is denoted by $\mathcal{FR}(X)$.

Remark 1.1 If $card\,X = n, X = \{x_1, \ldots, x_n\}, card\,Y = m, Y = \{y_1, \ldots, y_m\}$, then a fuzzy relation may be presented by a matrix $R = [r_{ik}]$, where $r_{ik} = R(x_i, y_k)$, $i = 1, \ldots, n, k = 1, \ldots, m, m, n \in \mathbb{N}$.

Example 1.1 Examples of fuzzy relations may be provided by expressions such as „x is almost equal to y", „x is about 10 times greater than y", „$x + y$ is close to 10", „x is much larger than y", „y is a good approximation of $\sqrt{x}$", „$x^2 + y^2$ is much smaller than 100", etc. Let $Y = X$, $card\ X = 5$. We present below an exemplary relation $R \in \mathcal{FR}(X)$ of *approximate equality* on a finite set X, where

$$R = \begin{bmatrix} 1 & 0.5 & 0.1 & 0 & 0 \\ 0.5 & 1 & 0.4 & 0.2 & 0 \\ 0.1 & 0.4 & 1 & 0.5 & 0.1 \\ 0 & 0.2 & 0.5 & 1 & 0.4 \\ 0 & 0 & 0.1 & 0.4 & 1 \end{bmatrix}.$$

Analogously to Definition 1.2 we may define basic operations on fuzzy relations. There exist also some operations that are typical for the relational calculus. One of such operations is composition. We recall the notion of sup–B and inf–B–composition. In these notions we follow the notational convention of Bandler and Kohout (cf. [10]).

Definition 1.4 (*cf.* [2]) Let $B : [0, 1]^2 \to [0, 1]$. A sup–B–composition of relations $R \in \mathcal{FR}(X, Y)$ and $W \in \mathcal{FR}(Y, Z)$ is the relation $(R \circ_B W) \in \mathcal{FR}(X, Z)$ such that for any $(x, z) \in X \times Z$ it holds

$$(R \circ_B W)(x, z) = \sup_{y \in X} B(R(x, y), W(y, z)). \tag{1.4}$$

An inf–B–composition of relations $R \in \mathcal{FR}(X, Y)$ and $W \in \mathcal{FR}(Y, Z)$ is the relation $(R \lhd_B W) \in \mathcal{FR}(X, Z)$ such that for any $(x, z) \in X \times Z$ it holds

$$(R \lhd_B W)(x, z) = \inf_{y \in X} B(R(x, y), W(y, z)). \tag{1.5}$$

In the next section the notions of fuzzy connectives are recalled. They are extensions of adequate notions of the classical propositional calculus.

1.1.2 Fuzzy Connectives

Fuzzy connectives are important notions in fuzzy sets theory. A fuzzy negation, a fuzzy conjunction, a fuzzy disjunction, a fuzzy implication, and a fuzzy equivalence are the basic connectives.

Definition 1.5 ([11]) A decreasing function $N : [0, 1] \to [0, 1]$ is called a fuzzy negation, if $N(0) = 1$ and $N(1) = 0$. A fuzzy negation is called a strict negation, if it is a strictly decreasing and continuous function. A fuzzy negation is called a strong negation, if it is involution, i.e. $N(N(x)) = x$ for any $x \in [0, 1]$.

Each strong fuzzy negation is a strict fuzzy negation but the converse is not true.

Example 1.2 (*cf.* [12]) Typical examples of fuzzy negations are:
• $N(x) = 1 - x$, which is a strong fuzzy negation and is called the classical or standard negation.
• $N(x) = 1 - x^2$, which is strict but not strong.
Moreover, the Sugeno family of fuzzy (strong) negations is the set of fuzzy negations of the form

$$N_S^{\lambda}(x) = \frac{1 - x}{1 + \lambda x}, \quad \lambda \in (-1, \infty), \quad x \in [0, 1].$$

Note that for $\lambda = 0$ we get the classical fuzzy negation.

The Yager family of strong fuzzy negations is the set of fuzzy negations of the form

$$N_Y^{w}(x) = (1 - x^{w})^{\frac{1}{w}}, \quad w \in (0, \infty), \quad x \in [0, 1].$$

The next class of fuzzy connectives consists of fuzzy conjunctions and disjunctions. There exist diverse definitions of these notions (cf. [12]). We recall one of the weakest approach to define fuzzy conjunctions and disjunctions along with some of their important subclasses.

Definition 1.6 (*cf.* [11, 13]) An operation $C : [0, 1]^2 \to [0, 1]$ is called a fuzzy conjunction (respectively disjunction) if it is increasing and $C(1, 1) = 1$, $C(0, 0) = C(0, 1) = C(1, 0) = 0$ (respectively $C(0, 0) = 0$, $C(1, 1) = C(0, 1) = C(1, 0) = 1$). A fuzzy conjunction is called a triangular seminorm (respectively triangular semiconorm) if it has a neutral element 1 (respectively 0). A triangular seminorm (respectively triangular semiconorm) is called a triangular norm (respectively triangular conorm) if it is commutative and associative.

Triangular norms (t-norms for short) and triangular conorms (t-conorms for short) are precisely discussed in the monograph [11].

Example 1.3 ([11]) Examples of fuzzy conjunctions are the following well-known t-norms (denoted by T) given here with their usually used abbreviations:
$T_M(x, y) = \min(x, y)$, $T_P(x, y) = xy$, $T_L(x, y) = \max(x + y - 1, 0)$,

$$T_D(x, y) = \begin{cases} x, & \text{if } y = 1 \\ y, & \text{if } x = 1 \\ 0, & \text{otherwise} \end{cases}.$$

Similarly, examples of fuzzy disjunctions are the following well-known t-conorms (denoted by S) given here with their usually used abbreviations:

$S_M(x, y) = \max(x, y)$, $S_P(x, y) = x + y - xy$, $S_L(x, y) = \min(x + y, 1)$,

$$S_D(x, y) = \begin{cases} x, & \text{if } y = 0 \\ y, & \text{if } x = 0 \\ 1, & \text{otherwise} \end{cases}.$$

These operations are comparable (linearly ordered on [0, 1]), i.e.

$$T_D \leqslant T_L \leqslant T_P \leqslant T_M, \quad S_D \geqslant S_L \geqslant S_P \geqslant S_M. \tag{1.6}$$

Other examples and classes of fuzzy conjunctions and disjunctions are gathered for example in [14].

Fuzzy implications are interesting connectives in fuzzy settings which may fulfil many additional properties (cf. [12, 15]).

Definition 1.7 ([15], *pp. 2, 9*) A function $I : [0, 1]^2 \to [0, 1]$ is called a fuzzy implication, if it is decreasing with respect to the first variable, increasing with respect to the second variable and fulfils the truth-table of a crisp implication, i.e. $I(1, 0) = 0$, $I(1, 1) = I(0, 1) = I(0, 0) = 1$.

A fuzzy implication I fulfils the identity principle, if

$$I(x, x) = 1, \quad x \in [0, 1]. \tag{1.7}$$

The notion of a fuzzy equivalence, understood as a fuzzy connective, is probably the least known among fuzzy connectives. However, this is important concept from practical point of view (cf. [16]) and it may be defined in diverse ways (cf. [17], p. 33, [18, 19]). We recall the notion of a fuzzy equivalence considered by Fodor and Roubens in [20].

Definition 1.8 ([20], *p. 33*) A function $E : [0, 1]^2 \to [0, 1]$ is called a fuzzy equivalence, if it fulfils the following conditions:

$$E(0, 1) = 0, \quad E(x, x) = 1, \quad E(x, y) = E(y, x), \quad x, y \in [0, 1],$$

$$E(x, y) \leqslant E(u, v), \quad x \leqslant u \leqslant v \leqslant y, \quad x, y, u, v \in [0, 1].$$

In Theorem 1.1 a generalization of a tautology from the classical propositional calculus, i.e. $(p \Leftrightarrow q) \Leftrightarrow ((p \Rightarrow q) \wedge (q \Rightarrow p))$, is applied to characterize a fuzzy equivalence given in Definition 1.8. However, the classical conjunction is replaced with a fuzzy conjunction – minimum.

Theorem 1.1 ([20], p. 33) *A function $E : [0, 1]^2 \to [0, 1]$ is a fuzzy equivalence if and only if there exists a fuzzy implication I fulfilling the identity principle* (1.7) *such that*

$$E(x, y) = \min(I(x, y), I(y, x)), \quad x, y \in [0, 1].$$

In the next section a brief overview of extensions of fuzzy sets are provided. Especially, some useful notions connected with interval-valued fuzzy sets and Atanassov intuitionistic fuzzy sets are recalled.

1.2 Elements of Interval-Valued Fuzzy Sets Theory

There are numerous extensions of fuzzy sets including one of the most important among them, i.e. interval-valued fuzzy sets [21, 22] introduced in 1970s and considered later by Gorzałczany [23] in 1987. Overview of the notions which extend the notion of a fuzzy set one may find in [24]. We briefly recall some of them. Hirota in 1977 introduced the notion of probabilistic sets [25]. Liu in 1982 considered for the first time the concept of a grey set [26]. In 1983 Atanassov presented at the conference the notion of intuitionistic fuzzy sets [27] and published the first paper on this subject [28] in 1986. This notion was suggested to be called an Atanassov intuitionistic fuzzy set due to some terminological difficulties [29]. Gau and Buehrer considered vague sets [30] in the seminal paper on this subject in 1993. One of the recent extensions of fuzzy sets are Pythagorean fuzzy sets [31] proposed by Yager in 2013. It is also worth to mention that Polish computer scientist Pawlak first described the concept of a rough set [32]. A rough set is a formal approximation of a crisp set in terms of a pair of sets which give the lower and the upper approximation of the original set. In the standard version of rough sets theory, the lower- and upper-approximation sets are crisp sets, but in other variations, the approximating sets may be fuzzy sets.

We concentrate here on the notion of an interval-valued fuzzy set and Atanassov intuitionistic fuzzy set. Interval-valued fuzzy sets are applied in many areas like databases, pattern recognition, neural networks, fuzzy modeling, economy, medicine or multicriteria decision making. Also in classification some algorithms for fuzzy rule based classification systems were improved using interval-valued fuzzy calculus. The use of intervals for the creation of the rules gives more flexibility to the algorithms, which leads to better results [33]. Intervals allow us to incorporate, the uncertainty or imprecision for instance in the original preference values given by the experts by means of the length of the intervals. The interval-valued fuzzy sets or relations may be obtained from fuzzy sets or relations, respectively using the concept of ignorance functions. This concept was defined in [34] to measure the degree of ignorance or lack of knowledge of an expert when he or she assigns numerical value as the membership degree of an object to a given class and another numerical value for the membership of the same element to a different class. Other construction methods of intervals from numbers and the application of the ignorance function to that number may be found for example in [35].

In this section we present the basic notions of interval-valued fuzzy calculus. We provide a discussion on diverse orders and comparability relations for intervals (not only for interval-valued fuzzy calculus) including the notions of linear orders for intervals. Moreover, we consider the notions of pos-B-transitivity, nec-B-transitivity and present information about interval-valued fuzzy connectives.

1.2.1 Basic Notions of Interval-Valued Fuzzy Calculus

Now, we recall definition of an interval-valued fuzzy set and the classically applied order for this settings. Throughout this book we apply the following notation

$$L^I = \{[\underline{x}, \overline{x}] : \underline{x}, \overline{x} \in [0, 1], \underline{x} \leqslant \overline{x}\}.$$

The well-known classical monotonicity (partial order) for L^I is of the form

$$[\underline{x}, \overline{x}] \preceq [\underline{y}, \overline{y}] \Leftrightarrow \underline{x} \leqslant \underline{y}, \ \overline{x} \leqslant \overline{y}. \tag{1.8}$$

Family L^I has bounds denoted by $\mathbf{0} = [0, 0]$, $\mathbf{1} = [1, 1]$ (cf. Fig. 1.1) and together with the relation $\preceq$ it is a complete and bounded lattice [36], where lattice operations are defined as follows

$$[\underline{x}, \overline{x}] \vee [\underline{y}, \overline{y}] = [\max(\underline{x}, \underline{y}), \max(\overline{x}, \overline{y})],$$

$$[\underline{x}, \overline{x}] \wedge [\underline{y}, \overline{y}] = [\min(\underline{x}, \underline{y}), \min(\overline{x}, \overline{y})].$$

In the interval-valued fuzzy sets theory to each element of the given universe a closed subinterval of the unit interval is assigned. This is the way of describing the unknown membership degree of the element to the given universe.

Definition 1.9 (*cf.* [22]) An interval-valued fuzzy set F on X is a mapping $F : X \to L^I$ such that $F(x) = [\underline{F}(x), \overline{F}(x)] \in L^I$ for $x \in X$, where $\underline{F}$ and $\overline{F}$ are fuzzy sets associated with the interval-valued fuzzy set F. The class of all interval-valued fuzzy sets on X is denoted by $\mathcal{IVFS}(X)$. The width $w_F : X \to [0, 1]$ of an interval valued fuzzy set $F(x) = [\underline{F}(x), \overline{F}(x)]$ is a function defined as

$$w_F(x) = \overline{F}(x) - \underline{F}(x), \quad x \in X. \tag{1.9}$$

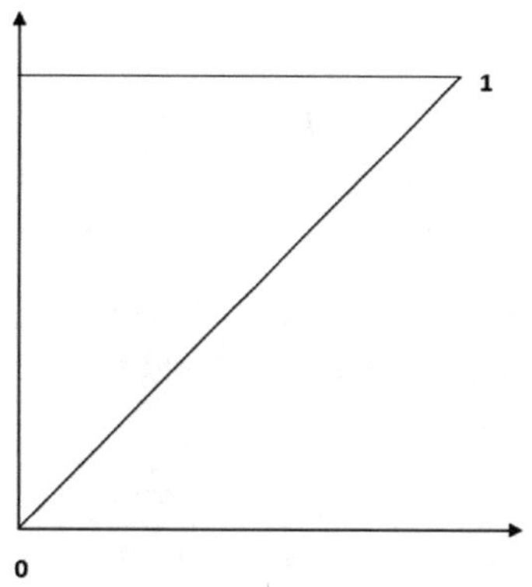

Fig. 1.1 Lattice L^I

The family $\mathcal{IVFS}(X)$ with the relation $\preceq$ is partially ordered and moreover it is a complete bounded lattice, which is a consequence of the fact that L^I with the order $\preceq$ is a complete and bounded lattice. Similar results may be obtained for interval-valued fuzzy relations.

Definition 1.10 (*cf.* [22]) An interval-valued fuzzy relation R between universes X, Y is a mapping $R : X \times Y \to L^I$ such that

$$R(x, y) = [\underline{R}(x, y), \overline{R}(x, y)] \in L^I,$$

for all couples $(x, y) \in (X \times Y)$, where $\underline{R}$ and $\overline{R}$ are fuzzy relations associated with the interval-valued fuzzy relation R. The class of all interval-valued fuzzy relations between universes X, Y is denoted by $\mathcal{IVFR}(X \times Y)$ or $\mathcal{IVFR}(X)$ for $X = Y$. $\mathcal{IVFR}(X \times Y)$ or $\mathcal{IVFR}(X)$ with the order $\preceq$ is a complete and bounded lattice.

It is worth to mention that interval-valued fuzzy sets are mathematically equivalent to Atanassov intuitionistic fuzzy sets (this is an isomorphism of structures). We will recall adequate theorem firstly giving suitable notions. Let us use the following notations

$$L^* = \{(x, y) : x, y \in [0, 1], \ x + y \leqslant 1\}$$

and the partial order for L^* be given by

$$(x_1, y_1) \preceq_{L^*} (x_2, y_2) \Leftrightarrow x_1 \leqslant x_2, \ y_2 \leqslant y_1. \tag{1.10}$$

Family L^* has the greatest element denoted by $\mathbf{1}_{L^*} = (1, 0)$ and the least element denoted by $\mathbf{0}_{L^*} = (0, 1)$ (cf. Fig. 1.2). Together with relation $\preceq_{L^*}$ it is a complete and bounded lattice, where lattice operations are defined as follows

$$(x_1, y_1) \vee (x_2, y_2) = (\max(x_1, x_2), \min(y_1, y_2)),$$

$$(x_1, y_1) \wedge (x_2, y_2) = (\min(x_1, x_2), \max(y_1, y_2)).$$

Definition 1.11 (*cf.* [28]) Let $X \neq \emptyset$, $\mu, \nu \colon X \to [0, 1]$ be fuzzy sets fulfilling the condition

$$\mu(x) + \nu(x) \leqslant 1, \qquad x \in X.$$

A pair $\rho = (\mu, \nu)$ is called an Atanassov intuitionistic fuzzy set. The family of all Atanassov intuitionistic fuzzy sets described on a given set X is denoted by $\mathcal{AIFS}(X)$.

The pair $(\mathcal{AIFS}(X), \preceq_{L^*})$ is a complete and bounded lattice. The isomorphism which proves the mathematical equivalence between Atanassov intuitionistic fuzzy sets and interval-valued fuzzy sets is the following.

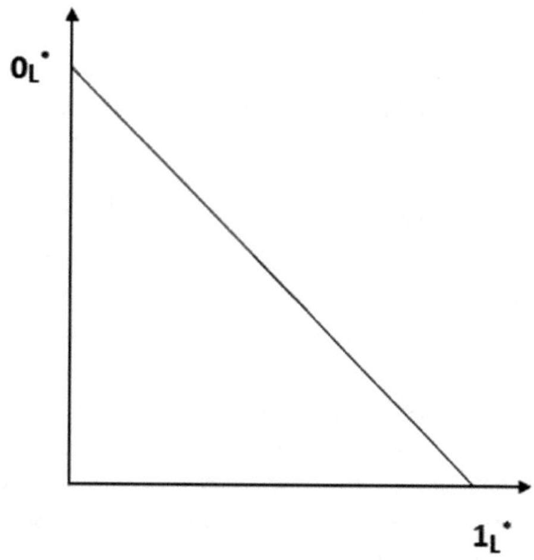

Fig. 1.2 Lattice L^*

Theorem 1.2 ([37], p. 139, [38, 39]) *A mapping* $\psi : \mathscr{AIFS}(X) \rightarrow \mathscr{IVFS}(X)$ *such that* $\rho \mapsto A$ *for all* $\rho = (\mu, \nu) \in \mathscr{AIFS}(X)$, *where* $A(x) = [\mu(x), 1 - \nu(x)]$, $x \in X$, *is an isomorphism.*

The existence of the membership degree interval $[\mu(x), 1 - \nu(x)]$ is always possible thanks to the condition $\mu(x) + \nu(x) \leqslant 1$, which each Atanassov intuitionistic fuzzy set $\rho = (\mu, \nu)$ should fulfill. The presented above isomorphism is due to the appropriate choice of operations in both theories, however the initial ideas behind each theory are different (cf. [29]). Interval-valued fuzzy set is defined by an interval-valued membership function representing the uncertainty or imprecision. Atanassov intuitionistic fuzzy set is defined by ascribing to an element a membership function and a non-membership function separately, in such a way that an element cannot have degrees of membership and non-membership that sum up to more than 1. The membership and non-membership degrees may represent the idea that concepts are more naturally approached by separately visualizing positive and negative information about the given fuzzy set.

The fuzzy set $\pi_\rho \colon X \rightarrow [0, 1]$ is associated with each Atanassov intuitionistic fuzzy set $\rho = (\mu, \nu)$, where

$$\pi_\rho(x) = 1 - \mu(x) - \nu(x), \quad x \in X.$$

The number $\pi_\rho(x)$ is called an index of an element x in ρ. This value is also regarded as a measure of non-determinacy or uncertainty (cf. [40]) and is useful in applications. The notion of π_ρ is equivalent of the notion of width of an interval-valued fuzzy set.

1.2.2 Order Relations for Interval-Valued Fuzzy Settings

In [41], a comparative study of the existing definitions of order relations between intervals, analyzing the level of acceptability and shortcomings from different points of view were presented. Some of them are used in interval-valued fuzzy settings. It is obvious that there is a need for effective defining relations of comparability, preferably orders and linear orders for interval-valued fuzzy settings. Moreover, the proposed definitions should have a reasonable interpretation. Comparability relations used in the interval-valued settings may be connected with *ontic* and *epistemic* interpretation of intervals [42, 43]. Epistemic uncertainty represents the idea of partial or incomplete information. Simply, it may be described by means of a set of possible values of some quantity of interest, one of which is the right one. A fuzzy set represents in such approach incomplete information, so it may be called disjunctive [42]. On the other hand, fuzzy sets may be conjunctive and can be called ontic fuzzy sets [42]. In this situation the fuzzy set is used as representing some precise gradual entity consisting of a collection of items.

In this monograph it is stressed the importance and usefulness in applications of the following two types of comparability relations for interval-valued fuzzy settings (cf. [44])

$$[\underline{x}, \overline{x}] \preceq_\pi [\underline{y}, \overline{y}] \Leftrightarrow \underline{x} \leqslant \overline{y}, \tag{1.11}$$

$$[\underline{x}, \overline{x}] \preceq_\nu [\underline{y}, \overline{y}] \Leftrightarrow \overline{x} \leqslant \underline{y}. \tag{1.12}$$

The origin of these relations is other than the one presented in [45] for partial order used classically in interval-valued fuzzy settings. These relations, including classical order, follow from the epistemic settings of interval-valued fuzzy sets (representing the uncertainty of membership value of a given object in an interval-valued fuzzy set, cf. [44]). Relation $\preceq_\pi$ between intervals may be interpreted as follows: at least one element in the first interval is smaller or equal to at least one element in the second interval. Relation $\preceq_\nu$ between intervals may be interpreted as follows: each element in the first interval is smaller or equal to each element in the second interval. Relation (1.8) (traditionally used in interval-valued settings) may be interpreted as a conjunction of the condition: there exists an element in the first interval that is smaller or equal to each element in the second interval and the condition: for each element in the first interval there exists a greater element in the second interval. Regarding this type of interpretation (all combinations of universal and existential quantifiers) by considering these three relations, i.e. $\preceq$, $\preceq_\pi$ and $\preceq_\nu$, we have the full possible set of interpretations of comparability relations on intervals regarding the epistemic settings. Let us note, that the relation (1.12) may also reflect the ontic approach of interval interpretation, which is not the case for the comparability relations (1.11) and (1.8).

To proceed with further properties of relations $\preceq_\pi$ and $\preceq_\nu$ we recall the definition of some classes of crisp relations that constitute order relations.

Definition 1.12 ([20], *p. 42*) A relation $R \subset X \times X$:

- is reflexive, if $R(x, x)$, for $x \in X$,
- is irreflexive, if $\sim R(x, x)$, for $x \in X$,
- is asymmetric, if $R(x, y) \Rightarrow \sim R(y, x)$, for $x, y \in X$,
- is antisymmetric, if $R(x, y)$ and $R(y, x) \Rightarrow x = y$, for $x, y \in X$,
- is complete, if $R(x, y)$ or $R(y, x)$, for $x, y \in X$,
- is transitive, if $R(x, y)$ and $R(y, z) \Rightarrow R(x, z)$, for $x, y, z \in X$,
- has Ferrers property, if $R(x, y)$ and $R(z, w) \Rightarrow R(x, w)$ or $R(z, y)$, for $x, y, z, w \in X$.

Relations $\preceq_\pi$ and $\preceq_\nu$ are not partial orders (reflexive, antisymmetric and transitive relations). Relation $\preceq_\pi$ is an interval order (complete and fulfils the Ferrers property, cf. [46–48]) and the relation $\preceq_\nu$ is antisymmetric and transitive.

In Fig. 1.3 exemplary intervals are illustrated by segments reflecting some of the possible cases of comparability between intervals $\mathbf{x} = [\underline{x}, \overline{x}]$, $\mathbf{y} = [\underline{y}, \overline{y}]$:
(a) according to relation (1.8), $\mathbf{x} \preceq \mathbf{y}$;
(b) according to relation (1.11), $\mathbf{x} \preceq_\pi \mathbf{y}$ and $\mathbf{y} \preceq_\pi \mathbf{x}$;
(c) according to relation (1.12), $\mathbf{x} \preceq_\nu \mathbf{y}$.

We see in Fig. 1.3 (case b)) that both $\mathbf{x} \preceq_\pi \mathbf{y}$ and $\mathbf{y} \preceq_\pi \mathbf{x}$. This may be problematic in practice, since using this comparability relation may require application of some additional steps in algorithms (cf. Sects. 3.2 and 6.2). Other notions of orders for intervals were introduced for example in the paper [45], where the general method to build different linear orders for the family of intervals L^I, covering some of the known linear orders for intervals was presented. This subject will be developed in the sequel part of the monograph.

It is worth to mention that other examples of orders for the family of intervals, considered not necessarily in the context of interval-valued fuzzy settings, were proposed for example in [41, 49–52]. Moore [51] developed an interval mathematics in

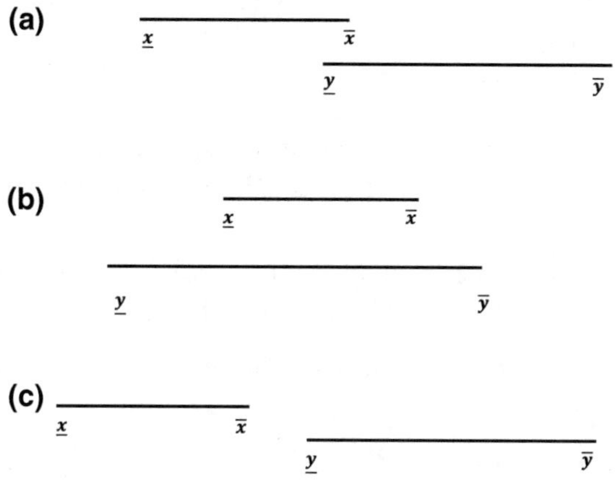

Fig. 1.3 Illustration of comparability relations on L^I

order to allow the control of errors in numeric computations in the solution of problems concerning real numbers. Moore first proposed two transitive order relations for intervals [51], where using the notations $\mathbf{x} = [\underline{x}, \overline{x}]$, $\mathbf{y} = [\underline{y}, \overline{y}]$, the following relations were defined:

- $\mathbf{x} \preceq_1 \mathbf{y} \Leftrightarrow \overline{x} < \underline{y}$,
- $\mathbf{x} \preceq_2 \mathbf{y} \Leftrightarrow (\underline{y} \leqslant \underline{x}$ and $\overline{x} \leqslant \overline{y})$.

Relation $\preceq_1$, which is a strict order relation (irreflexive, asymmetric and transitive relation), was proposed for non-overlapping intervals. Relation $\preceq_2$ is the generalization of the definition of subsets for intervals and it follows the properties of partial order relation as the classical set inclusion operation $\subset$ possesses.

In [50] were presented a few relations of comparability between intervals which are considered in the rough set approach to multicriteria classification along with the corresponding P-dominance relations. P-dominance relation is connected with the dominance principle which requires that an object having not worse evaluations on condition criteria than another object should be assigned to the class not worse than the other object. The imprecision in classifications are given in the form of intervals of possible values. These are the following relations:

- $\mathbf{x} \preceq_3 \mathbf{y} \Leftrightarrow \underline{x} \leqslant \overline{y}$,
- $\mathbf{x} \preceq_4 \mathbf{y} \Leftrightarrow \underline{x} \leqslant \underline{y}$,
- $\mathbf{x} \preceq_5 \mathbf{y} \Leftrightarrow \overline{x} \leqslant \overline{y}$.

Relation $\preceq_3$ is equal to the relation (1.11), relations $\preceq_4$ and $\preceq_5$ are complete quasi-orders (reflexive, transitive and complete relations).

The order (1.8) was first defined for the family of closed intervals of the real line $I(\mathbb{R})$ (it is called the Kulish-Miranker order [53, 54]). Despite this order there exist also some other orders for the family $I(\mathbb{R})$, namely:

$\mathbf{x} \preceq_6 \mathbf{y} \Leftrightarrow (\mathbf{x} = \mathbf{y}$ or $\overline{x} \leqslant \underline{y})$ (the Moore order, cf. [55, 56]),
$\mathbf{x} \preceq_7 \mathbf{y} \Leftrightarrow (\underline{x} \leqslant \underline{y} \leqslant \overline{y} \leqslant \overline{x})$ (the information order, cf. [52, 57]).

Many other orders for intervals have been developed to serve various purposes (cf. [58–63]). Detailed survey for these definitions were given in [41, 64] with their advantages and disadvantages. The primary goal of these definitions is to develop reliable solution technique for interval optimization problems with the help of interval ranking.

Below there will be presented more information about comparability relations $\preceq_\nu$ and $\preceq_\pi$, since in this monograph we concentrate on the results connected with these notions.

Proposition 1.1 (cf. [44, 65]) *Relation $\preceq_\pi$ is an interval order and the relation $\preceq_\nu$ is antisymmetric and transitive on L^I. Moreover, $\preceq_\nu \Rightarrow \preceq \Rightarrow \preceq_\pi$.*

The converse implications do not hold, as can be seen from the following example.

Example 1.4 For intervals $A = [0.2, 0.3]$, $B = [0.1, 0.5]$ and $C = [0.4, 0.8]$ we observe that $A \preceq_\pi B$ but it is not true that $A \preceq B$ and similarly $B \preceq C$ but it is not true that $B \preceq_\nu C$.

Proposition 1.2 (cf. [65]) *The boundary elements on L^I, with respect to each of the comparability relations* (1.8), (1.11), (1.12), *are* $\mathbf{1} = [1, 1]$ *and* $\mathbf{0} = [0, 0]$.

Relations $\preceq_\pi$ and $\preceq_\nu$ may be defined also on the family $\mathcal{IVFS}(X)$ (and certainly $\mathcal{IVFR}(X, Y)$). However, on such domain they may posses another set of properties than they have on L^I.

Proposition 1.3 (cf. [44, 65]) *The relation $\preceq_\pi$ on $\mathcal{IVFS}(X)$ is reflexive and the relation $\preceq_\nu$ on $\mathcal{IVFS}(X)$ is antisymmetric and transitive.*

Note that, relations (1.8), (1.11) and (1.12) coincide on the family of fuzzy sets. It means that if the given $A, B \in \mathcal{IVFS}(X)$ are such that for some $x \in X$ their membership intervals $[\underline{A}(x), \overline{A}(x)]$ and $[\underline{B}(x), \overline{B}(x)]$ fulfil the condition $\underline{A}(x) = \overline{A}(x)$ and $\underline{B}(x) = \overline{B}(x)$, then applying each of the relations (1.8), (1.11) and (1.12) to the given intervals, we obtain the order relation for fuzzy sets (1.2).

In the next section we provide the results referring to the notion of an *admissible linear order*, which was defined for interval-valued fuzzy settings.

1.2.3 Linear Orders for Interval-Valued Fuzzy Settings

The classical monotonicity (partial order), which is traditionally used for intervals, is the consequence of the following interpretation. To each interval $[a, b] \in L^I$ one may assign uniquely a point $(a, b) \in \mathbb{R}^2$, so intervals (thanks to isomorphism) may be ordered by means of orders for points on $\mathbb{R}^2$. The usual partial order on $\mathbb{R}^2$, given by $(a, b) \preceq (c, d) \Leftrightarrow a \leqslant c$ and $b \leqslant d$, induces the partial order (1.8) (cf. [45]). Classical order (1.8) is not complete (which is important for application reasons for example in decision making problems to compare alternatives and choose the best one). It is worth to recall that in the paper [45] the general method to build different linear orders, i.e. complete order relations covering some of the known linear orders for intervals, such as Xu and Yager's [66] or the lexicographical ones, is presented.

Definition 1.13 ([45]) An order $\leq_{L^I}$ on L^I is called admissible if it is linear and satisfies that for all $x, y \in L^I$, such that $x \preceq y$, then $x \leq_{L^I} y$.

According to the above definition it means that an order $\leq_{L^I}$ on L^I is admissible, if it is linear and refines the order $\preceq$. There exist a construction method of admissible linear orders with the use of aggregation functions (an aggregation function $A : [0, 1]^2 \to [0, 1]$ is an increasing in each variable function fulfilling boundary conditions $A(0, 0) = 0$ and $A(1, 1) = 1$, more detailed information on the concept of an aggregation functions will be given in Chap. 2).

Proposition 1.4 ([45]) *Let $B_1, B_2 : [0, 1]^2 \to [0, 1]$ be two continuous aggregation functions, such that, for all $x = [\underline{x}, \overline{x}]$, $y = [\underline{y}, \overline{y}] \in L^I$, the equalities $B_1(\underline{x}, \overline{x}) = B_1(\underline{y}, \overline{y})$ and $B_2(\underline{x}, \overline{x}) = B_2(\underline{y}, \overline{y})$ hold if and only if $x = y$. If the order $\leq_{B_{1,2}}$ on L^I is defined by $x \leq_{B_{1,2}} y$ if and only if*

$$B_1(\underline{x}, \overline{x}) < B_1(\underline{y}, \overline{y}) \ or \ (B_1(\underline{x}, \overline{x}) = B_1(\underline{y}, \overline{y}) \ and \ B_2(\underline{x}, \overline{x}) \leqslant B_2(\underline{y}, \overline{y})),$$

then $\leq_{B_{1,2}}$ *is an admissible order on* L^I.

The presented construction method is based on the application of two aggregation functions. In this sense, it provides an easy-to-use tool in order to generate new linear orders between intervals (depending on the requirements of the real-life situations). Moreover, the presented method can be easily extended to use either more aggregation functions or functions other than aggregation functions as long as some basic properties are fulfilled (cf. [45]).

Example 1.5 Well-known examples of admissible (linear) orders on L^I are presented below:

- Xu and Yager order:
  $[\underline{x}, \overline{x}] \leq_{XY} [\underline{y}, \overline{y}]$ if and only if
  $$\underline{x} + \overline{x} < \underline{y} + \overline{y} \ or \ (\overline{x} + \underline{x} = \overline{y} + \underline{y} \ and \ \overline{x} - \underline{x} \leqslant \overline{y} - \underline{y}).$$
- The first lexicographical order (with respect to the first variable), $\leq_{Lex1}$ defined as: $[\underline{x}, \overline{x}] \leq_{Lex1} [\underline{y}, \overline{y}]$ if and only if
  $$\underline{x} < \underline{y} \ or \ (\underline{x} = \underline{y} \ and \ \overline{x} \leqslant \overline{y}).$$
- The second lexicographical order (with respect to the second variable), $\leq_{Lex2}$ defined as:
  $[\underline{x}, \overline{x}] \leq_{Lex2} [\underline{y}, \overline{y}]$ if and only if
  $$\overline{x} < \overline{y} \ or \ (\overline{x} = \overline{y} \ and \ \underline{x} \leqslant \underline{y}).$$
- Let $K_\alpha : [0, 1]^2 \to [0, 1]$ be the function defined as $K_\alpha(x, y) = \alpha x + (1 - \alpha)y$ for some $\alpha \in [0, 1]$. The order defined as $x \leq_{\alpha\beta} y$ if and only if
  $$K_\alpha(\underline{x}, \overline{x}) < K_\alpha(\underline{y}, \overline{y}) \ or \ (K_\alpha(\underline{x}, \overline{x}) = K_\alpha(\underline{y}, \overline{y}) \ and \ K_\beta(\underline{x}, \overline{x}) \leqslant K_\beta(\underline{y}, \overline{y}))$$
  is an admissible order for $\alpha \neq \beta$ and $x, y \in L^I$.

As it may be noticed, the orders $\leq_{XY}$, $\leq_{Lex1}$, $\leq_{Lex2}$ and $\leq_{\alpha\beta}$ are special cases of the order $\leq_{B_{1,2}}$. Linear orders may be also constructed with the use of other operators [67]. We present below an example of such construction. There exist several operators designed to act on intervals, for example Atanassov's operators defined in [37] and studied in [68].

Let $[\underline{x}, \overline{x}] \in L^I, \alpha, \beta \in [0, 1]$ with $\alpha + \beta \leqslant 1$. The so called Atanassov's operator is defined as $F_{\alpha,\beta} : L^I \to L^I$ with

$$F_{\alpha,\beta}([\underline{x}, \overline{x}]) = [\underline{x} + \alpha(\overline{x} - \underline{x}), \overline{x} - \beta(\overline{x} - \underline{x})]. \tag{1.13}$$

A particular way of obtaining admissible orders is defining them by means of $F_{\alpha,1-\alpha}$ mappings. Note that if $\alpha + \beta = 1$, then by (1.13) we have

$$F_{\alpha,1-\alpha}([\underline{x}, \overline{x}]) = [\underline{x} + \alpha(\overline{x} - \underline{x}), \underline{x} + \alpha(\overline{x} - \underline{x})] = \underline{x} + \alpha(\overline{x} - \underline{x}),$$

thus $F_{\alpha,1-\alpha}$ is a weighted mean.

Let $\alpha,\ \beta \in [0, 1]$, $\alpha \neq \beta$. The relation $\preceq_{\alpha,\beta}$ on L^I given by

$$\begin{aligned}[\underline{x}, \overline{x}] \preceq_{\alpha,\beta} [\underline{y}, \overline{y}] \ &\Leftrightarrow F_{\alpha,1-\alpha}(\underline{x}, \overline{x}) < F_{\alpha,1-\alpha}(\underline{y}, \overline{y}) \\ &\text{or } (F_{\alpha,1-\alpha}(\underline{x}, \overline{x}) = F_{\alpha,1-\alpha}(\underline{y}, \overline{y}) \text{ and } F_{\beta,1-\beta}(\underline{x}, \overline{x}) \leqslant F_{\beta,1-\beta}(\underline{y}, \overline{y})), \quad (1.14)\end{aligned}$$

is an admissible order generated by an admissible pair of aggregation functions $(F_{\alpha,1-\alpha}, F_{\beta,1-\beta})$.

The $F_{\alpha,\beta}$ operators where widely studied in the context of interval-valued fuzzy relations and preference modeling in [69]. In particular, in that work the following result was proved.

Proposition 1.5 (cf. [69]) *Let* $R \in \mathscr{IVFR}(X)$, $\alpha, \beta \in [0, 1]$ *with* $\alpha + \beta \leqslant 1$. *Then, for every* $x, y \in X$

$$\underline{R}(x, y) \leqslant \underline{F_{\alpha,\beta}(R)}(x, y) \leqslant \overline{F_{\alpha,\beta}(R)}(x, y) \leqslant \overline{R}(x, y). \quad (1.15)$$

The above property means that the operator $F_{\alpha,\beta}(R)$ applied to R makes the amplitude of the interval of R smaller, so in the interpretation in preference modelling the uncertainty is smaller. Moreover, the iterative application of Atanassov's operator over a given interval-valued fuzzy relation eventually reduces it to a fuzzy relation.

1.2.4 Possible *and* Necessary *Properties of Interval-Valued Fuzzy Relations*

If it comes to the comparability relations $\preceq_\pi$ and $\preceq_\nu$, which are one of the topic of interest in this monograph, there may be considered some new notions for interval-valued fuzzy settings involving the mentioned comparability relations. For example new types of properties of interval-valued fuzzy relations. The novelty follows from the fact of replacing classical order $\preceq$ with the relations $\preceq_\pi$ and $\preceq_\nu$. Below there is presented the general rule of creating such new properties.

Definition 1.14 (*cf.* [44, 65]) Relation $R \in \mathscr{IVFR}(X)$ has possible p–property (pos-p–property for short) if there exists at least one instance R^* of R which has property p. Relation $R \in \mathscr{IVFR}(X)$ has necessary p–property (nec-p–property for short) if for every instance R^* of R it has property p.

One of the most important properties is transitivity. This is why we present here only this concept in the new version, i.e. with the relations $\preceq_\pi$ and $\preceq_\nu$ involved and generalized to the form depending on a binary operation B (cf. [14]).

Definition 1.15 (*cf.* [44]) Let $B : [0, 1]^2 \to [0, 1]$ be a binary operation. A relation $R = [\underline{R}, \overline{R}] \in \mathcal{IVFR}(X)$ is:

- B-transitive, if

$$B(\underline{R}(x, y), \underline{R}(y, z)) \leqslant \underline{R}(x, z) \text{ and } B(\overline{R}(x, y), \overline{R}(y, z)) \leqslant \overline{R}(x, z), \quad x, y, z \in X,$$

- necessarily B-transitive (nec-B-transitive), if

$$B(\overline{R}(x, y), \overline{R}(y, z)) \leqslant \underline{R}(x, z), \quad x, y, z \in X,$$

- possibly B-transitive (pos-B-transitive), if

$$B(\underline{R}(x, y), \underline{R}(y, z)) \leqslant \overline{R}(x, z), \quad x, y, z \in X.$$

In Definition 1.15 an arbitrary operation $B : [0, 1]^2 \to [0, 1]$ is applied in order to obtain the most general version of the properties. However, it is natural to consider there a fuzzy conjunction B, which generalizes the operation of minimum used in the classical definition of transitivity. This approach is also a generalization of the crisp meaning of transitivity (cf. Definition 1.12). Similarly to pos-B-transitivity and nec-B-transitivity we may also define other properties.

We present below some dependencies between transitivity properties given in Definition 1.15. Similar dependencies were provided in [44] but only for pos-B-transitivity and B-transitivity in the special case $B = T$, where T is a triangular norm. The notions of B-transitivity, pos-B-transitivity and nec-B-transitivity may be characterized with the use of composition. Since these properties are expressed with the use of lower and upper bounds of membership intervals, then the characterization is provided with the use of the notion of sup–B–composition for fuzzy relations.

Proposition 1.6 *Let $R \in \mathcal{IVFR}(X)$, $B : [0, 1]^2 \to [0, 1]$.*

- *R is B-transitive if and only if $\underline{R} \circ_B \underline{R} \leqslant \underline{R}$ and $\overline{R} \circ_B \overline{R} \leqslant \overline{R}$.*
- *R is nec-B-transitive if and only if $\overline{R} \circ_B \overline{R} \leqslant \underline{R}$.*
- *R is pos-B-transitive if and only if $\underline{R} \circ_B \underline{R} \leqslant \overline{R}$.*

Proof R is pos-B-transitive if and only if

$$\forall_{x,y,z\in X} \; B(\underline{R}(x, y), \underline{R}(y, z)) \leqslant \overline{R}(x, z) \Leftrightarrow \forall_{x,z\in X} \; \sup_{y\in X} B(\underline{R}(x, y), \underline{R}(y, z)) \leqslant \overline{R}(x, z) \Leftrightarrow$$

$$\forall_{x,z\in X} \; (\underline{R} \circ_B \underline{R})(x, z) \leqslant \overline{R}(x, z) \Leftrightarrow \underline{R} \circ_B \underline{R} \leqslant \overline{R}.$$

The justification for nec-B-transitivity and B-transitivity is analogous.

Proposition 1.7 *Let* $R \in \mathcal{IVFR}(X)$, $B : [0, 1]^2 \to [0, 1]$.

- *If* R *is nec-*B*-transitive and* B *is increasing, then* R *is* B*-transitive.*
- *If* R *is* B*-transitive, then* R *is pos-*B*-transitive.*

Proof By Proposition 1.6 if R is B-transitive, then $\underline{R} \circ_B \underline{R} \leqslant \underline{R} \leqslant \overline{R}$, so we immediately obtain the second property. If R is nec-B-transitive, then by Proposition 1.6 we get $\overline{R} \circ_B \overline{R} \leqslant \underline{R} \leqslant \overline{R}$, so we get $\overline{R} \circ_B \overline{R} \leqslant \overline{R}$. Next, by monotonicity of B we have $\underline{R} \circ_B \underline{R} \leqslant \overline{R} \circ_B \overline{R} \leqslant \underline{R}$, which means that $\underline{R} \circ_B \underline{R} \leqslant \underline{R}$. Finally, we see that R is B-transitive, which finishes the proof of the second property.

Connections between diverse types of B-dependent transitivity properties with respect to diverse operations B are given below.

Proposition 1.8 ([70]) *Let* $B_1, B_2 : [0, 1]^2 \to [0, 1]$ *and* $B_1 \leqslant B_2$. *If* $R \in \mathcal{IVFR}(X)$ *is pos–*B_2*–transitive (nec–*B_2*–transitive,* B_2*–transitive), then* R *is pos–*B_1*–transitive (nec–*B_1*–transitive,* B_1*–transitive).*

As a result we have monotonicity in the family of properties with respect to the applied operations B.

1.2.5 Interval-Valued Fuzzy Connectives

When we are working on interval-valued fuzzy settings an opportunity to consider many types of connectives arises. Namely, we may for example consider diverse representations of such notions. Below we recall the concept of a representable interval-valued fuzzy negation. Other presentations, like for example the so called best interval representation, one may find in [71].

Definition 1.16 (*cf.* [72]) $\mathcal{N}$ is called a representable interval-valued fuzzy negation if there exists a fuzzy negation N such that $\mathcal{N}([\underline{x}, \overline{x}]) = [N(\overline{x}), N(\underline{x})]$, where $[\underline{x}, \overline{x}] \in L^I$. We say that N is the associate fuzzy negation with the given representable interval-valued fuzzy negation.

The following result links strong representable interval-valued fuzzy negations with strong fuzzy negations.

Theorem 1.3 ([72]) $\mathcal{N}$ *is a strong representable interval-valued fuzzy negation if and only if there exists a strong fuzzy negation* N *such that* $\mathcal{N}([\underline{x}, \overline{x}]) = [N(\overline{x}), N(\underline{x})]$, *where* $[\underline{x}, \overline{x}] \in L^I$.

An important notion which involves the concept of an interval-valued fuzzy negation is the notion of a $\mathcal{N}$-dual operator to the given one, which will be considered in Chap. 2. Interval-valued fuzzy settings gives us also the possibility to use diverse notions of orders. Interval-valued fuzzy negations with respect to admissible linear orders were considered in [73, 74].

Notions of other connectives such as disjunctions and disjunctions for interval-valued fuzzy settings one may find in [75, 76]. In particular, the manners to construct different interval-valued fuzzy connectives (or Atanassovs intuitionistic fuzzy connectives) starting from an operator are provided.

Diverse types of interval-valued fuzzy implications are examined in [77–80]. Moreover, interval-valued fuzzy implications with respect to admissible linear orders were introduced and considered in [74]. Equivalence for interval-valued fuzzy settings were presented, along with applications, for example in [81]. Generally speaking, one of the possibilities to define an interval-valued connective is to create an analogous notion to the one existing on the unit interval [0, 1]. However, we may create many versions of a given connective using diverse representations and diverse types of orders.

An important notion for interval-valued fuzzy settings is the notion of an aggregation operator. In [82], the class of linear orders on L^I was used to extend the definition of OWA operators to interval-valued fuzzy settings. This definition as well as other examples, construction methods and properties of the aggregation operators will be recalled in the next part of the monograph (Chap. 2).

References

1. Zadeh, L.A.: Fuzzy logic-a personal perspective. Fuzzy Sets Syst. **281**, 4–20 (2015)
2. Zadeh, L.A.: Fuzzy sets. Inf. Control **8**, 338–353 (1965)
3. Klaua, D.: Über einen Ansatz zur mehrwertigen Mengenlehre. Monatsb. Deutsch. Akad. Wiss. Berlin **7**, 859–876 (1965) A recent in-depth analysis of this paper has been provided by Gottwald, S.: An early approach toward graded identity and graded membership in set theory. Fuzzy Sets Syst. **161**(18), 2369–2379 (2010)
4. Łukasiewicz, J: O logice trójwartościowej (in Polish). Ruch filozoficzny **5**, 170–171 (1920) English translation: On three-valued logic. In: Borkowski L. (eds.) Selected works by Jan Łukasiewicz, pp. 87–88. North Holland, Amsterdam (1970)
5. Szpilrajn, E.: The characteristic function of a sequence of sets and some of its applications. Fund. Math. **31**, 207–223 (1938)
6. Menger, K.: Ensembles flous et fonctions aléatoires. C. R. Acad. Sci. Paris **232**, 2001–2003 (1951)
7. Rasiowa, H.: A generalization of a formalized theory of fields of sets on non-classical logics. Rozpr. Matemat. **42**, 3–29 (1964)
8. Zadeh, L.A.: Similarity relations and fuzzy orderings. Inf. Sci. **3**, 177–200 (1971)
9. Goguen, A.: L-fuzzy sets. J. Math. Anal. Appl. **18**, 145–174 (1967)
10. Bandler, W., Kohout, L.J.: Semantics of implication operators and fuzzy relational products. Int. J. Man-Mach. Stud. **12**, 89–116 (1980)
11. Klement, E.P., Mesiar, R., Pap, E.: Triangular Norms. Kluwer Academic Publishers, Dordrecht (2000)
12. Pradera, A., Beliakov, G., Bustince, H., De Baets, B.: A review of the relationships between implication, negation and aggregation functions from the point of view of material implication. Inf. Sci. **329**, 357–380 (2016)
13. Drewniak, J., Król, A.: A survey of weak connectives and the preservation of their properties by aggregations. Fuzzy Sets Syst. **161**, 202–215 (2010)
14. Bentkowska, U., Król, A.: Preservation of fuzzy relation properties based on fuzzy conjunctions and disjunctions during aggregation process. Fuzzy Sets Syst. **291**, 98–113 (2016)

15. Baczyński, M., Jayaram, B.: Fuzzy Implications. Studies in Fuzziness and Soft Computing, vol. 231. Springer, Berlin (2008)
16. Bustince, H., Barrenechea, E., Pagola, M.: Image thresholding using restricted equivalence functions and maximizing the measures of similarity. Fuzzy Sets Syst. **158**, 496–516 (2007)
17. Nguyen, H.T., Walker, E.: A First Course in Fuzzy Logic. CRC Press, Boca Raton (1996)
18. Bustince, H., Barrenechea, E., Pagola, M.: Restricted equivalence functions. Fuzzy Sets Syst. **157**, 2333–2346 (2006)
19. Bentkowska, U., Król, A.: Fuzzy α-C-equivalences. Fuzzy Sets Syst. (2018). https://doi.org/10.1016/j.fss.2018.01.004
20. Fodor, J., Roubens, M.: Fuzzy Preference Modelling and Multicriteria Decision Support. Kluwer Academic Publisher, Dordrecht (1994)
21. Sambuc, R.: Fonctions ϕ-floues: Application á l'aide au Diagnostic en Pathologie Thyroidienne. Ph.D. thesis, Universit*é* de Marseille, France (1975) (in French)
22. Zadeh, L.A.: The concept of a linguistic variable and its application to approximate reasoning-I. Inf. Sci. **8**, 199–249 (1975)
23. Gorzałczany, M.B.: A method of inference in approximate reasoning based on interval-valued fuzzy sets. Fuzzy Sets Syst. **21**, 1–17 (1987)
24. Bustince, H., Barrenechea, E., Pagola, M., Fernandez, J., Xu, Z., Bedregal, B., Montero, J., Hagras, H., Herrera, F., De Baets, B.: A historical account of types of fuzzy sets and their relationships. IEEE Trans. Fuzzy Syst. **24**(1), 179–194 (2016)
25. Hirota, K.: Concept of probabilistic sets. In: Proceedings of IEEE Conference on Decision and Control, pp. 1361–1366. New Orleans (1977)
26. Liu, K.: Grey sets and stability of grey systems. J. Huazhong Univ. Sci. Technol. **10**(3), 23–25 (1982)
27. Atanassov, K.T.: Intuitionistic fuzzy sets. In: Proceedings of VII ITKRs Session, pp. 1684–1697. Sofia, Bulgaria (1983)
28. Atanassov, K.T.: Intuitionistic fuzzy sets. Fuzzy Sets Syst. **20**, 87–96 (1986)
29. Dubois, D., Gottwald, S., Hajek, P., Kacprzyk, J., Prade, H.: Terminological difficulties in fuzzy set theory - the case of intuitionistic fuzzy sets. Fuzzy Sets Syst. **156**, 485–491 (2005)
30. Gau, W.L., Buehrer, D.J.: Vague sets. IEEE Trans. Syst. Man Cybern. **23**(2), 610–614 (1993)
31. Yager, R.R.: Pythagorean fuzzy subsets. In: Proceedings of the Joint IFSA World Congress and NAFIPS Annual Meeting, pp. 57–61 (2013)
32. Pawlak, Z.: Rough sets. Int. J. Comput. Inf. Sci. **11**, 341–356 (1982)
33. Sanz, J., Fernandez, A., Bustince, H., Herrera, F.: A genetic tuning to improve the performance of fuzzy rule-based classification systems with intervalvalued fuzzy sets: degree of ignorance and lateral position. Int. J. Approx. Reason. **52**(6), 751–766 (2011)
34. Bustince, H., Pagola, M., Barrenechea, E., Fernandez, J., Melo-Pinto, P., Couto, P., Tizhoosh, H.R., Montero, J.: Ignorance functions. An application to the calculation of the threshold in prostate ultrasound images. Fuzzy Sets Syst. **161**(1), 20–36 (2010)
35. Barrenechea, E., Fernandez, J., Pagola, M., Chiclana, F., Bustince, H.: Construction of interval-valued fuzzy preference relations from ignorance functions and fuzzy preference relations: application to decision making. Knowl. Based Syst. **58**, 33–44 (2014)
36. Birkhoff, G.: Lattice Theory. AMS Colloquium Publications XXV, Providence (1967)
37. Atanassov, K.T.: Intuitionistic Fuzzy Sets: Theory and Applications. Springer, Berlin (1999)
38. Deschrijver, G., Kerre, E.E.: On the relationship between some extensions of fuzzy set thory. Fuzzy Sets Syst. **133**(2), 227–235 (2003)
39. Deschrijver, G., Kerre, E.E.: On the position of intuitionistic fuzzy set theory in the framework of theories modelling imprecision. Inf. Sci. **177**, 1860–1866 (2007)
40. Lin, L., Yuan, X.-H., Xia, Z.-Q.: Multicriteria fuzzy decision-making methods based on intuitionistic fuzzy sets. J. Comput. Syst. Sci. **73**, 84–88 (2007)
41. Karmakar, S., Bhunia, A.K.: A comparative study of different order relations of intervals. Reliab. Comput. **16**, 38–72 (2012)
42. Dubois, D., Prade, H.: Gradualness, uncertainty and bipolarity: making sense of fuzzy sets. Fuzzy Sets Syst. **192**, 3–24 (2012)

43. Dubois, D., Godo, L., Prade, H.: Weighted logics for artificial intelligence an introductory discussion. Int. J. Approx. Reason. **55**, 1819–1829 (2014)
44. Pękala, B., Bentkowska, U., De Baets, B.: On comparability relations in the class of interval-valued fuzzy relations. Tatra Mt. Math. Publ. **66**, 91–101 (2016)
45. Bustince, H., Fernandez, J., Kolesárová, A., Mesiar, R.: Generation of linear orders for intervals by means of aggregation functions. Fuzzy Sets Syst. **220**, 69–77 (2013)
46. Fishburn, P.C.: Intransitive indifference with unequal indifference intervals. J. Math. Psychol. **7**, 144–149 (1970)
47. Fishburn, P.C.: Utility Theory for Decision Making. Wiley, New York (1970)
48. Fishburn, P.C.: Interval Orders and Interval Graphs. Wiley, New York (1985)
49. Callejas-Bedregal, R., Callejas Bedregal, B.R.: Intervals as a domain constructor. TEMA - Tendências em Matemática Aplicada e Computacional **2**(1), 43–52 (2001)
50. Dembczyński, K., Greco, S., Sowiński, R.: Rough set approach to multiple criteria classification with imprecise evaluations and assignments. Eur. J. Oper. Res. **198**, 626–636 (2009)
51. Moore, R.E.: Interval Analysis, vol. 4. Prentice-Hall, Englewood Cliffs (1966)
52. Scot, D.S.: Outline of a mathematical theory of computation. In: 4th Annual Princeton Conference on Information Sciences and Systems, pp. 169–176 (1970)
53. Kulish, U.W., Miranker, W.L.: Computer Arithmetic in Theory and Practice. Technical report 33658, IBM Thomas L. Watson Research Center (1979)
54. Kulish, U.W., Miranker, W.L.: Computer Arithmetic in Theory and Practice. Academic, New York (1981)
55. Moore, R.E.: Methods and Applications for Interval Analysis. SIAM, Philadelfia (1979)
56. Dimuro, G.P., Costa, A.C.R., Claudio, D.M.: A coherent space of rational intervals for construction of IFR. J. Rielable Comput. **6**, 139–178 (2000)
57. Acióly, B.M.: Computational Foundation of Interval Mathematic. Ph.D. thesis (in Portugeese). CPGCC, UFRGS, Porto Allegre (1991)
58. Sengupta, A., Pal, T.K.: On comparing interval numbers. Eur. J. Oper. Res. **127**(1), 28–43 (2000)
59. Ishibuchi, H., Tanaka, H.: Multiobjective programming in optimization of the interval objective function. Eur. J. Oper. Res. **48**(2), 219–225 (1990)
60. Chanas, S., Kuchta, D.: Multiobjective programming in optimization of interval objective functions - a generalized approach. Eur. J. Oper. Res. **94**(3), 594–598 (1996)
61. Mahato, S.K., Bhunia, A.K.: Interval-arithmetic-oriented interval computing technique for global optimization. Appl. Math. Res. Express **1–19**, (2006)
62. Moore, R.E., Kearfott, R.B., Cloud, M.J.: Introduction to Interval Analysis. SIAM, Philadelphia (2009)
63. Karmakar, S., Bhunia, A.K.: An alternative optimization technique for interval objective constrained optimization problems via multiobjective programming. J. Egypt. Math. Soc. **22**, 292–303 (2014)
64. Sengupta, A., Pal, T.K.: Fuzzy Preference Ordering of Interval Numbers in Decision Problems. Springer, Berlin (2009)
65. Pękala, B.: Uncertainty Data in Interval-Valued Fuzzy Set Theory. Properties, Algorithms and Applications. Studies in Fuzziness and Soft Computing. Springer, Cham, Switzerland (2019)
66. Xu, Z.S., Yager, R.R.: Some geometric aggregation operators based on intuitionistic fuzzy sets. Int. J. Gen. Syst. **35**, 417–433 (2006)
67. Bentkowska, U., Bustince, H., Jurio, A., Pagola, M., Pękala, B.: Decision making with an interval-valued fuzzy preference relation and admissible orders. Appl. Soft Comput. **35**, 792–801 (2015)
68. Bustince, H.: Construction of intuitionistic fuzzy sets with predetermined properties. Fuzzy Sets Syst. **109**, 379–403 (2000)
69. Bustince, H., Burillo, P.: Perturbation of intuitionistic fuzzy relations. Int. J. Uncertain. Fuzziness Knowl. Based Syst. **9**, 81–103 (2001)
70. Bentkowska, U.: New types of aggregation functions for interval-valued fuzzy setting and preservation of pos-B and nec-B-transitivity in decision making problems. Inf. Sci. **424**, 385–399 (2018)

71. Bedregal, B.: On interval fuzzy negations. Fuzzy Sets Syst. **161**(17), 2290–2313 (2010)
72. Deschrijver, G., Cornelis, C., Kerre, E.: On the representation of intuitonistic fuzzy t-norms and t-conorms. IEEE Trans. Fuzzy Syst. **12**, 45–61 (2004)
73. Asiaín, M.J., Bustince, H., Mesiar, R., Kolesárová, A., Takáč, Z.: Negations with respect to admissible orders in the interval-valued fuzzy set theory. IEEE Trans. Fuzzy Syst. **26**(2), 556–568 (2018)
74. Zapata, H., Bustince, H., Montes, S., Bedregal, B., Dimuro, G.P., Takáč, Z., Baczyński, M., Fernandez, J.: Interval-valued implications and interval-valued strong equality index with admissible orders. Int. J. Approx. Reason. **88**, 91–109 (2017)
75. Bustince, H., Montero, J., Pagola, M., Barrenechea, E., Gomez, D.: A survey of interval-valued fuzzy sets. In: Pedrycz, W., Skowron, A., Kreinovich, V. (eds.) Handbook of Granular Computing, pp. 489–515. Wiley, New York (2008)
76. Bustince, H., Barrenechea, E., Pagola, M.: Generation of interval-valued fuzzy and atanassovs intuitionistic fuzzy connectives from fuzzy connectives and from K_α operators: laws for conjunctions and disjunctions, amplitude. Int. J. Intell. Syst. **23**, 680–714 (2008)
77. Bedregal, B., Dimuro, G., Santiago, R., Reiser, R.: An approach to interval-valued R-implications and automorphisms. In: Carvalho, J.P., Dubois, D., Kaymak, U., Sousa, J.M.C. (eds.) Proceedings of the Joint 2009 International Fuzzy Systems Association World Congress and 2009 European Society of Fuzzy Logic and Technology Conference, Lisbon, Portugal, pp. 1–6. ISBN: 978-989-95079-6-8 (20–24 July, 2009)
78. Bedregal, B., Dimuro, G., Santiago, R., Reiser, R.: On interval fuzzy S-implications. Inf. Sci. **180**(8), 1373–1389 (2010)
79. Cornelis, C., Deschrijver, G., Kerre, E.E.: Implication in intuitionistic fuzzy and interval-valued fuzzy set theory: construction, classification, application. Int. J. Approx. Reason. **35**(1), 55–95 (2004)
80. Reiser, R.H.S., Dimuro, G.P., Bedregal, B.C., Santiago, R.H.N.: Interval valued QL-implications. In: Leivant D., De Queiroz R. (eds.) Logic, Language, Information and Computation. WoLLIC 2007. Lecture Notes in Computer Science, vol. 4576, pp. 307–321. Springer, Berlin (2007)
81. Jurio, A., Pagola, M., Paternain, D., Lopez-Molina, C., Melo-Pinto, P.: Interval-valued restricted equivalence functions applied on clustering technique. In: Carvalho, J.P., Dubois, D., Kaymak, U., Sousa, J.M.C. (eds.) Proceedings of the Joint 2009 International Fuzzy Systems Association World Congress and 2009 European Society of Fuzzy Logic and Technology Conference, Lisbon, Portugal, pp. 831–836. ISBN: 978-989-95079-6-8 (20–24 July, 2009)
82. Bustince, H., Galar, M., Bedregal, B., Kolesárová, A., Mesiar, R.: A new approach to interval-valued Choquet integrals and the problem of ordering in interval-valued fuzzy sets applications. IEEE Trans. Fuzzy Syst. **21**(6), 1150–1162 (2013)

Chapter 2
Aggregation in Interval-Valued Settings

My crystal ball is fuzzy

Lotfi A. Zadeh

Aggregation functions for interval-valued fuzzy settings derive from the concept of an aggregation function defined on the unit interval $[0, 1]$. This is why we begin with recalling the most important properties of aggregation functions as well as a short historical development of this concept. Next, classes, properties and construction methods of aggregation functions defined for interval-valued settings are provided. We pay special attention to the new classes of aggregation operators, namely possible and necessary aggregation functions. We present examples, properties, construction methods and dependencies between these two classes of operators and other well-known aggregation operators applied in interval-valued fuzzy settings. The presented results may be also useful for the entire community, not only fuzzy, involved in research under uncertainty or imperfect information.

2.1 Aggregation Functions

This section presents the concept of an aggregation function, the historical development of this notion, basic properties and the most important classes. Moreover, in the last part of the section the notion of dominance, as an interesting and useful concept concerning aggregation functions, is recalled.

U. Bentkowska, *Interval-Valued Methods in Classifications and Decisions*,
Studies in Fuzziness and Soft Computing 378,
https://doi.org/10.1007/978-3-030-12927-9_2

2.1.1 Development of the Concept of Aggregation Function

Well-known examples of aggregation functions are means, including the arithmetic mean

$$M(x_1, \ldots, x_n) = \frac{x_1 + \cdots + x_n}{n}. \tag{2.1}$$

To most non mathematicians this is the only way of averaging a set of numbers. The arithmetic mean (cf. [1, 2], p. 34) of two values (natural numbers or linear segments) occurred in the works of the Pythagorean school, sixth century B.C. For example, in the concept of arithmetic proportion of natural numbers we have $(a - b) : (b - c) = 1$, where $b = (a + c) : 2$. The same idea is in Archimedes' concept of centroid, third century B.C. Two other means, i.e. geometric mean

$$M(x_1, \ldots, x_n) = \sqrt[n]{x_1 \cdot \ldots \cdot x_n}, \;\; x_i \geqslant 0, \; i = 1, \ldots, n \tag{2.2}$$

and harmonic mean

$$M(x_1, \ldots, x_n) = \frac{n}{\frac{1}{x_1} + \cdots + \frac{1}{x_n}}, \quad x_i > 0, \; i = 1, \ldots, n \tag{2.3}$$

are fairly well-known. Like the arithmetic mean they arose naturally in Euclid as solutions to simple, algebraic and geometric problems. For example in the Pythagorean school's theory over harmony one may find the proportion

$$a : \frac{a+b}{2} = \frac{2ab}{a+b} : b,$$

where $\frac{2ab}{a+b} = 2 : (\frac{1}{a} + \frac{1}{b})$ is the harmonic mean of two values (named by Hippas, fourth century B.C.). There are three basic approaches to the most important definitions in the theory of means, i.e. functional, approximational and axiomatic (cf. [3]). Here there is presented the axiomatic approach. Since the XIX century a mean has been considered as an n-argument function fulfilling some axioms. One of the first authors who considered the concept of a mean was Cauchy in 1821.

Definition 2.1 ([4]) A mean of n independent variables $x_1, \ldots, x_n$ is a function $M(x_1, \ldots, x_n)$ which is internal, i.e.

$$\min(x_1, \ldots, x_n) \leqslant M(x_1, \ldots, x_n) \leqslant \max(x_1, \ldots, x_n). \tag{2.4}$$

The inequality (2.4) is also called a compensation property. Chisini in 1929 generalized the notion of mean in the following way.

Definition 2.2 ([5], *p. 108*) A mean of variables $x_1, \ldots, x_n$, with respect to an n-argument function g, is a number M such that

$$g(M, \ldots, M) = g(x_1, \ldots, x_n). \tag{2.5}$$

Example 2.1 Let $g(x_1, \ldots, x_n) = x_1 + \cdots + x_n$, then $g(M) = nM$. As a result the solution of the equation (2.5) with respect to the function M is of the form (2.1). Similarly, when g is the product, the sum of squares, the sum of inverses or the sum of exponentials, then the solution of Chisini's equation corresponds to the geometric, the quadratic, the harmonic or the exponential mean, respectively (cf. Example 2.3).

In 1931 Finetti ([6], p. 378) has observed that M from (2.5) may not fulfil (2.4).

Example 2.2 ([7], *p. 39*) Let $g(x_1, \ldots, x_n) = x_n + (x_n - x_1) + \cdots + (x_n - x_{n-1})$, then for $x_n > x_i$, $i = 1, \ldots, n-1$ we have $M = g(M, \ldots, M) = g(x_1, \ldots, x_n) > x_n$, which is contradictory to (2.4).

In 1930 Kolmogorov [8] and Nagumo [9] introduced, independently of each other, the following definition of a mean.

Definition 2.3 ([8]) A mean value is a sequence $(M_n)_{n\in\mathbb{N}}$ of functions $M_n : [a, b]^n \to [a, b]$, where $[a, b] \subset \mathbb{R}$, which fulfil the following conditions:

$$\underset{n\in\mathbb{N}}{\forall} \; M_n \text{ is continuous,} \tag{2.6}$$

$$\underset{n\in\mathbb{N}}{\forall} \; M_n(x_1, \ldots, x_n) = M_n(x_{\alpha(1)}, \ldots, x_{\alpha(n)}) \text{ for all permutations } \alpha = (\alpha(1), \ldots, \alpha(n)) \tag{2.7}$$

of the set $(1, \ldots, n)$ – M_n is symmetric,

$$\underset{n\in\mathbb{N}}{\forall} \; M_n \text{ is strictly increasing with respect to any variable,} \tag{2.8}$$

$$\underset{n\in\mathbb{N}}{\forall} \; \underset{x\in[a,b]}{\forall} \; M_n(x, \ldots, x) = x \; - M_n \text{ is idempotent,} \tag{2.9}$$

$$\underset{n\in\mathbb{N}}{\forall} \; \underset{k\leqslant n}{\forall} \; \underset{x_1,\ldots,x_n\in[a,b]}{\forall} \; M_n(x_1, \ldots, x_k, x_{k+1}, \ldots, x_n) = M_n(x, \ldots, x, x_{k+1}, \ldots, x_n), \tag{2.10}$$

where $x = M_k(x_1, \ldots, x_k)$.

Kolmogorov and Nagumo are also the authors of the following fundamental result.

Theorem 2.1 ([8]) *Conditions* (2.6)–(2.10) *are necessary and sufficient for the existence of a continuous and strictly monotonic function* $f : [a, b] \to \mathbb{R}$ *such that*

$$M_n(x_1, \ldots, x_n) = f^{-1}\left(\frac{1}{n}\sum_{i=1}^{n} f(x_i)\right), \quad n \in \mathbb{N}. \tag{2.11}$$

If a function f defines a mean (2.11), then a function $g(x) = \alpha f(x) + \beta$, $\alpha \neq 0$, $\beta \in \mathbb{R}$, defines the same mean. As a consequence of (2.11) $M_n \in \mathbb{N}$ is isomorphic with the arithmetic mean and it is called the *quasi-arithmetic mean*.

Finetti [6] and Kitagawa [10] extended the concept of the quasi-arithmetic mean by considering the positive weights $w_1, \ldots, w_n$ such that $\sum_{i=1}^{n} w_i = 1$, instead of $w_i = \frac{1}{n}$ in (2.11). Functions obtained in this way are of the form

$$M_n(x_1, \ldots, x_n) = f^{-1}\left(\sum_{i=1}^{n} w_i f(x_i)\right), \quad w_i > 0, \sum_{i=1}^{n} w_i = 1, \; n \in \mathbb{N}. \tag{2.12}$$

They are called *quasi-linear means* (cf. [11], p. 287). They are not symmetric so the characterization theorem is of the following form.

Theorem 2.2 ([6]) *Conditions* (2.6), (2.8)–(2.10) *are necessary and sufficient for the existence of a continuous and strictly monotonic function* $f : [a, b] \to \mathbb{R}$ *such that* (2.12) *holds.*

Many other authors dealt with the problem of a characterization of the quasi-arithmetic and quasi-linear means, e.g. Aczél [12], Aczél and Alsina [13], Aczél and Dhombres [14], Fodor and Marichal [15], Marichal [1].

Example 2.3 (*cf.* [1]) Let $n \in \mathbb{N}$, $w_k > 0$, $\sum_{k=1}^{n} w_k = 1$. Well-known quasi-linear means on [0, 1] are listed in Table 2.1.

For $w_k = \frac{1}{n}$, $k = 1, \ldots, n$ we get respective quasi-arithmetic means. Moreover,

$$P_{(1)} = A, \quad P_{(2)} = Q, \quad P_{(-1)} = H,$$

$$\lim_{r \to 0} P_{(r)} = G, \quad \lim_{r \to -\infty} P_{(r)} = \min, \quad \lim_{r \to \infty} P_{(r)} = \max .$$

There exist some well-known inequalities involving quasi-linear means.

Theorem 2.3 (cf. [16], p. 26) *Using assumptions and notions of Example 2.3, with the fixed weights, the following inequalities hold:*

$$\underset{x_1,\ldots,x_n \in [0,1]}{\forall} \; H(x_1, \ldots, x_n) \leqslant G(x_1, \ldots, x_n) \leqslant A(x_1, \ldots, x_n) \leqslant Q(x_1, \ldots, x_n), \tag{2.13}$$

$$\underset{r<s}{\forall} \quad \underset{x_1,\ldots,x_n \in [0,1]}{\forall} \quad P_{(r)}(x_1, \ldots, x_n) \leqslant P_{(s)}(x_1, \ldots, x_n). \tag{2.14}$$

The inequality (2.14) refers also to the limit cases of power means (cf. Example 2.3). Many theoretical aspects of quasi-linear means may be found in the monograph of Hardy, Littlewood and Pólya [16]. There exist several generalizations of these functions. For example in the papers of Bajraktarevič [17] or Páles [18] as solutions of the functional equations we may have the following means

$$M_{f,g}(x_1, \ldots, x_n) = f^{-1}\left(\sum_{i=1}^{n} g(x_i) f(x_i) / \sum_{i=1}^{n} g(x_i)\right), \tag{2.15}$$

Table 2.1 Types of weighted means

no.	$\varphi(x)$	weighted mean	type
1.	x	$A(x_1,\ldots,x_n)=\sum_{k=1}^{n} w_k x_k$	arithmetic
2.	x^2	$Q(x_1,\ldots,x_n)=\sqrt{\sum_{k=1}^{n} w_k x_k^2}$	quadratic
3.	$\log x$	$G(x_1,\ldots,x_n)=\prod_{k=1}^{n} x_k^{w_k}$	geometric
4.	x^{-1}	$H(x_1,\ldots,x_n)=\begin{cases}0, & \exists_{1\leqslant k\leqslant n}\ x_k=0\\ (\sum_{k=1}^{n}\frac{w_k}{x_k})^{-1}, & \text{otherwise}\end{cases}$	harmonic
5.	$x^r,\ r\in\mathbb{R}_0$	$P_{(r)}(x_1,\ldots,x_n)=\begin{cases}0, & r<0,\ \exists_{1\leqslant k\leqslant n}\ x_k=0\\ (\sum_{k=1}^{n} w_k x_k^r)^{\frac{1}{r}}, & \text{otherwise}\end{cases}$	power
6.	$e^{rt}, r\in\mathbb{R}_0$	$E(x_1,\ldots,x_n)=\frac{1}{r}\ln(\sum_{k=1}^{n} w_k e^{rx_k})$	exponential

$$M_{f,g}(x_1,\ldots,x_n)=f^{-1}\left(\sum_{i=1}^{n} w_i g(x_i) f(x_i)/\sum_{i=1}^{n} w_i g(x_i)\right), \tag{2.16}$$

where $f:[a,b]\to\mathbb{R}$ is a continuous, strictly increasing function and $g:[a,b]\to\mathbb{R}$ is a function of positive values. Function (2.15) is called a quasi-mixture function (cf. [19]). We also have other types of mixture functions and as a special case of (2.15) we obtain the Gini means (cf. [20]). If $g(x_i)=p$ is a constant then in (2.15) we have the quasi-arithmetic mean and in (2.16) the quasi-linear mean. It is worth mentioning that the arithmetic, geometric and harmonic means have their integral analogues, for aggregation of infinitely many values (cf. [21]), respectively given by:

$$A(f)=\frac{1}{b-a}\int_a^b f(t)dt,$$

$$G(f)=\exp\left(\frac{1}{b-a}\int_a^b ln\ f(t)dt\right),$$

$$H(f)=\frac{b-a}{\int_a^b \frac{dt}{f(t)}},$$

where f is a positive function defined on $[a, b]$. In the case of a step function f the integral changes into a sum and we get suitable discrete versions.

Example 2.4 Other types of means than the quasi-linear ones are the family of *Lehmer means* (cf. [2], p. 185)

$$L_{(r)}(x_1, \ldots, x_n) = \frac{\sum\limits_{k=1}^{n} w_k x_k^r}{\sum\limits_{k=1}^{n} w_k x_k^{r-1}}$$

where $w_k > 0$, $\sum_{k=1}^{n} w_k = 1$, $r \in \mathbb{R}$, and there is used the convention $\frac{0}{0} = 0$. For the weights $w_k = \frac{1}{n}$, $k = 1, \ldots, n$ one obtains the means of respective types. Let us note that Lehmer means are a special case of Gini means. Moreover, the following properties hold

$$L_{(1)} = A, \quad L_{(0)} = H,$$

$$L_{(\frac{1}{2})} = G \quad \text{in the case} \quad n = 2,$$

$$\lim_{r \to -\infty} P_{(r)} = \min, \quad \lim_{r \to \infty} P_{(r)} = \max,$$

$$L_{(2)}(x_1, \ldots, x_n) = \frac{\sum\limits_{k=1}^{n} w_k x_k^2}{\sum\limits_{k=1}^{n} w_k x_k}, \qquad \text{with the convention } \frac{0}{0} = 0,$$

where $L_{(2)}$ is the most known example of the family of Lehmer means and it is called just the Lehmer mean or the *contraharmonic mean*. The following monotonicity condition describes the family of Lehmer means

$$\forall_{r,s \in \mathbb{R}} \; r \leqslant s \Rightarrow L_{(r)} \leqslant L_{(s)}. \tag{2.17}$$

Moreover, one has the following dependencies between families $L_{(r)}$ and and described in Example 2.3 $P_{(r)}$ (cf. [2], p. 186)

$$r \leqslant 1 \Rightarrow L_{(r)} \leqslant P_{(r)}, \tag{2.18}$$

$$r \geqslant 1 \Rightarrow P_{(r)} \leqslant L_{(r)}. \tag{2.19}$$

In the eighties of the XX century some ordered means on $[0, 1]$ were introduced.

Definition 2.4 ([22]) $M : [0, 1]^n \to [0, 1]$ is said to be an ordered weighted averaging operator (OWA for short) if there exists a sequence of weights $w = (w_1, \ldots, w_n) \in [0, 1]^n$, $\sum\limits_{i=1}^{n} w_i = 1$, such that

$$\underset{(x_1,\ldots,x_n)\in[0,1]^n}{\forall} \quad M(x_1,\ldots,x_n)=\sum_{i=1}^{n} w_i y_i,$$

where $(y_1,\ldots,y_n)$, $y_1 \geqslant y_2 \geqslant \cdots \geqslant y_n$ is a sequence of arguments $(x_1,\ldots,x_n)$ ordered decreasingly.

Some examples of OWA operators are provided below.

Example 2.5 If $w=(0,0,\ldots,0,1)$, then $M(x_1,\ldots,x_n)=\min(y_1,\ldots,y_n)$.
If $w=(1,0,\ldots,0,0)$, then $M(x_1,\ldots,x_n)=\max(y_1,\ldots,y_n)$.
If $w=(\frac{1}{n},\ldots,\frac{1}{n})$, then $M(x_1,\ldots,x_n)=\frac{y_1+\cdots+y_n}{n}$.
If n is odd and $w=(0,0,\ldots,1,\ldots,0,0)$, then $M(x_1,\ldots,x_n)=y_{(n+1)/2}$,
if n is even and $w=(0,0,\ldots,\frac{1}{2},\frac{1}{2},\ldots,0,0)$, then $M(x_1,\ldots,x_n)=\frac{y_{n/2}+y_{(n/2)+1}}{2}$,
which together gives the median function

$$\operatorname{med}(x_1,\ldots,x_n)=\begin{cases} y_{(n+1)/2}, & \text{if } n \text{ is odd} \\ \frac{y_{n/2}+y_{(n/2)+1}}{2}, & \text{if } n \text{ is even} \end{cases}. \tag{2.20}$$

Furthermore, the so-called *olympic aggregation* obtained by assigning $w_1=w_n=0$ and $w_i=\frac{1}{n-2}$ for all other weights is an OWA operator. Here there are essentially disregarded the highest and the lowest scores.

The characterization of OWA operators was given by Fodor, Marichal and Roubens in 1995.

Theorem 2.4 ([23]) *The class of OWA operators corresponds to the operators which are symmetric, increasing, idempotent functions fulfilling condition*

$$\underset{x_1\leqslant\ldots\leqslant x_n\in[0,1]}{\forall} \ \underset{t_1\leqslant\ldots\leqslant t_n\in[0,1]}{\forall} \ \underset{r>0}{\forall} \ M(rx_1+t_1,\ldots,rx_n+t_n)=rM(x_1,\ldots,x_n)+M(t_1,\ldots,t_n), \tag{2.21}$$

where $rx_i+t_i\in[0,1]$ *for* $i=1,\ldots,n$.

Definition of a mean given by Kolmogorov has been weakened through the years in order to obtain a wider family of means. In addition there have been made some changes in the matter of the name of a *mean* which was replaced with *aggregation function* or *averaging function*. As a result conditions (2.6), (2.7), (2.10) from Definition 2.3 were omitted. This lead to the following definition, where strict monotonicity was replaced with the monotonicity condition.

Definition 2.5 (*cf.* [24], *p. 107*) Let $n\geqslant 2$. $M:\mathbb{R}^n\to\mathbb{R}$ is said to be an aggregation function if it is increasing and idempotent, i.e.

$$\underset{x_1,\ldots,x_n,y_1,\ldots,y_n\in\mathbb{R}}{\forall} \left(\left(\underset{1\leqslant i\leqslant n}{\forall} x_i\leqslant y_i\right)\Rightarrow M(x_1,\ldots,x_n)\leqslant M(y_1,\ldots,y_n)\right), \tag{2.22}$$

$$\underset{x\in\mathbb{R}}{\forall} M(x,\ldots,x)=x. \tag{2.23}$$

Condition (2.22) means that if one input increases while the others are kept constant, the overall degree of M must not decrease. An arbitrary aggregation function (cf. Definition 2.5) fulfils the compensation property (2.4) and as a consequence it may be considered in any closed interval. For the convenience and because of the application reasons (in fuzzy sets theory and its extensions) we take into account the unit interval [0, 1]. Examples of aggregation functions according to Definition 2.5 are quasi-arithmetic means, quasi-linear means or OWA operators. However, Lehmer means do not fulfil condition (2.22), so they are not aggregation functions according to Definition 2.5.

The next step of enlarging the family of aggregation functions is done by weakening the condition (2.23).

Definition 2.6 (*cf.* [25], *Definition 1*) Let $n \geqslant 2$. $A : [0, 1]^n \to [0, 1]$ is said to be an aggregation function if it fulfils (2.22) and boundary conditions

$$A(0, \ldots, 0) = 0, \quad A(1, \ldots, 1) = 1. \tag{2.24}$$

The condition (2.24) requires idempotency only for the boundary elements of the interval in which the aggregation function is defined. This fact enables to obtain new families of aggregation functions. Aggregation functions according to Definition 2.6 are fuzzy conjunctions and disjunctions presented in Definition 1.6. Certainly, each function fulfilling the axioms of Definition 2.5 fulfils also the axioms of Definition 2.6.

The family of all aggregation functions on [0, 1] is bounded.

Proposition 2.1 ([25]) *Aggregation functions* $A_w, A_s : [0, 1]^n \to [0, 1]$ *are the lower and upper bounds of the family of all aggregation functions on* [0, 1], *respectively*

$$A_w(x_1, \ldots, x_n) = \begin{cases} 1, & (x_1, \ldots, x_n) = (1, \ldots, 1) \\ 0, & \textit{otherwise}, \end{cases}$$

$$A_s(x_1, \ldots, x_n) = \begin{cases} 0, & (x_1, \ldots, x_n) = (0, \ldots, 0) \\ 1, & \textit{otherwise}, \end{cases}$$

A_w *is called the weakest aggregation function on* [0, 1] *and* A_s *is called the strongest aggregation function on* [0, 1].

Remark 2.1 In the sequel, if using the notion of an aggregation function, we will consider Definition 2.6. This definition is now commonly used when regarding aggregation functions. Similarly, the letter A will be used to denote aggregation functions fulfilling axioms of Definition 2.6.

Means given by Bajraktarevič (with special case of Gini means) or mixture functions admit many averaging functions (i.e. functions fulfilling the condition (2.4)) that are not aggregation functions since they are not monotone in the sense of (2.22).

Lehmer means are one of the examples of averaging functions that are not monotone in the sense of (2.22), so they are not aggregation functions. More information about other types of monotonicity and new classes of aggregation functions will be provided in Sect. 2.1.2.

2.1.2 Classes of Aggregation Function

There have been mentioned so far some axioms for means or aggregation functions, i.e. (2.4), (2.6)–(2.10), (2.21). We recall some other important ones which may be also defined for arbitrary operation on [0, 1] (not necessarily an aggregation function).

Definition 2.7 (*cf.* [25]) Let $n \in \mathbb{N}$. An aggregation function $A\colon [0,1]^n \to [0,1]$: is bisymmetric, if for an arbitrary matrix $[x_{ik}] \in [0,1]^{n\times n}$ the following equality holds

$$A(A(x_{11},\ldots,x_{1n}),\ldots,A(x_{n1},\ldots,x_{nn})) = A(A(x_{11},\ldots,x_{n1}),\ldots,A(x_{1n},\ldots,x_{nn})), \tag{2.25}$$

has a neutral element $e \in [0,1]$, if for any $k \in \{1,\ldots,n\}$

$$\underset{x_1,\ldots,x_{k-1},x_k,x_{k+1},\ldots,x_n\in[0,1]}{\forall} A(x_1,\ldots,x_{k-1},e,x_{k+1},\ldots,x_n) = A(x_1,\ldots,x_{k-1},x_{k+1},\ldots,x_n), \tag{2.26}$$

has a zero element $z \in [0,1]$ (called also *annihilator*), if

$$\underset{1\leqslant k\leqslant n}{\forall}\ \underset{x_1,\ldots,x_{k-1},x_{k+1},\ldots,x_n\in[0,1]}{\forall} A(x_1,\ldots,x_{k-1},z,x_{k+1},\ldots,x_n) = z, \tag{2.27}$$

is without zero divisors if it has a zero element z and

$$\underset{x_1,\ldots,x_n\in[0,1]}{\forall} (A(x_1,\ldots,x_n) = z \Rightarrow (\underset{1\leqslant k\leqslant n}{\exists} x_k = z)). \tag{2.28}$$

Some additional conditions for an aggregation function (cf. Definition 2.6) are considered for example in [24, 25]. Aggregation functions may be divided into four classes.

Definition 2.8 ([26]) Aggregation function A is called:

- conjunctive, if $A \leqslant \min$,
- disjunctive, if $A \geqslant \max$,
- averaging, if $\min \leqslant A \leqslant \max$,
- hybrid (or mixed), if they are neither conjunctive nor disjunctive nor averaging.

The above mentioned four classes of aggregation functions constitute the partition of the whole set of aggregation functions. Examples of conjunctive and disjunctive aggregation functions are conjunctions and disjunctions, respectively. They have variety of subclasses (cf. Definition 1.6). Examples of averaging aggregation functions are given in Example 2.3. Examples of hybrid aggregation functions are binary operations such as uninorms (associative and symmetric aggregation functions with a neutral element $e \in (0, 1)$, cf. [27]) and nullnorms (associative and symmetric aggregation functions that possess an element $a \in (0, 1)$ such that for any $x \in [0, 1]$, it holds $A(x, 0) = x$ if $x \leqslant a$ and $A(x, 1) = x$ if $x \geqslant a$, cf. [28]). A more detailed classification of aggregation functions, regarding their important subfamilies, one may find in [26].

There are also considered other than (2.22) types of monotonicity (increasingness) for aggregation functions $A : [0, 1]^n \to [0, 1]$.

Definition 2.9 ([25]) An aggregation function $A : [0, 1]^n \to [0, 1]$ is called jointly strictly monotone (increasing), if for all $x_1, \ldots, x_n, y_1, \ldots, y_n \in [0, 1]$

$$(\forall_{1 \leqslant i \leqslant n} x_i < y_i) \Rightarrow A(x_1, \ldots, x_n) < A(y_1, \ldots, y_n). \tag{2.29}$$

Example 2.6 ([25]) Aggregation functions which are jointly strictly monotone are: quasi-linear means, min, max, product $A(x_1, \ldots, x_n) = \prod_{i=1}^{n} x_i$.

Definition 2.10 ([20], *p. 11*) An aggregation function $A : [0, 1]^n \to [0, 1]$ is called strictly monotone (increasing), if for all $x = (x_1, \ldots, x_n)$, $y = (y_1, \ldots, y_n) \in [0, 1]^n$

$$(x \leqslant y \text{ and } x \neq y) \Rightarrow A(x) < A(y). \tag{2.30}$$

Example 2.7 (*cf.* [20], *p. 12*) Aggregation functions which are strictly monotone on [0, 1] are quasi-linear means (cf. Definition 2.3). Product $A(x_1, \ldots, x_n) = \prod_{i=1}^{n} x_i$ is strictly increasing on (0, 1]. Minimum and maximum are not strictly increasing. Let us consider $x = (0.6, 0.6)$, $y = (0.8, 0.6)$, then coordinate-wise $x \leqslant y$ and $x \neq y$ but $\min(x_1, x_2) = \min(y_1, y_2) = 0.6$. Similarly, we may show that max is not strictly increasing. As a result there are no strictly increasing conjunctive (or disjunctive) aggregation functions on [0, 1]. This holds because every conjunctive function coincides with $\min(x)$ for those x that have at least one component x_i equal to zero. Similarly, every disjunctive function coincides with $\max(x)$ for those x that have at least one component x_i equal to one. However, as we have given an example of the product function, strict monotonicity of aggregation functions may be considered on the semi-open interval (0, 1] in the case of conjunctive aggregation functions and on the semi-open interval [0, 1) in the case of disjunctive aggregation functions.

Now, we recall the notion of a dual function which is often useful when describing properties of subclasses of aggregation functions and also other operations on [0, 1]. One of such classes consist of binary aggregation functions—conjunctions and their dual operations disjunctions. These notions (including some of their subclasses) were recalled in Chap. 1.

Definition 2.11 (*cf.* [25], *p. 31*) Let $F : [0,1]^n \to [0,1]$. A function F^d is called a dual function to F, if for all $x_1, \ldots, x_n \in [0,1]$

$$F^d(x_1, \ldots, x_n) = 1 - F(1 - x_1, \ldots, 1 - x_n).$$

F is called a self-dual function, if it holds $F = F^d$.

Due to the fact that the dual functions to fuzzy conjunctions are fuzzy disjunctions and vice versa, and these two classes are disjoint it follows that neither a fuzzy conjunction nor a fuzzy disjunction (including t-norms and t-conorms) is a self-dual function. For any binary function F, if it is without zero divisors (with the zero element $z = 0$), then F cannot be self-dual (cf. [29]). Any self-dual and symmetric binary aggregation function F satisfies $F(x, 1 - x) = \frac{1}{2}$ for all $x \in [0,1]$. The concept of self-duality is especially developed for aggregation functions. Interesting properties and characterizations of self-dual aggregation functions one can find in [30]. For example, we recall the result involving quasi-linear means.

Example 2.8 ([30]) A weighted arithmetic mean, median and all quasi-linear means for which $f : [0,1] \to [0,1]$ in (2.11) fulfils $f(1 - x) = 1 - f(x)$, are self-dual aggregation functions.

The notion of a dual function may be generalized to the notion of a N-dual function with respect to an arbitrary strong fuzzy negation N.

Definition 2.12 (*cf.* [31]) Let $F : [0,1]^n \to [0,1]$. A function F^N is called an N-dual function to F, if for all $x_1, \ldots, x_n \in [0,1]$ and some strong fuzzy negation N it holds

$$F^N(x_1, \ldots, x_n) = N(F(N(x_1), \ldots, N(x_n))).$$

F is called a self N-dual function, if it holds $F = F^N$.

In Definition 2.11, the function F^d is defined with respect to a classical fuzzy negation $N(x) = 1 - x$.

Nowadays, the notion of an aggregation function is weakened by considering weaker versions of monotonicity instead of the monotonicity condition (2.22). Recently, the directional monotonicity condition with respect to some vector $\overrightarrow{r}$ was introduced (cf. [32]). This resulted in introducing a weaker version of the concept of an aggregation function, namely a *pre-aggregation function*.

Definition 2.13 ([33]) A function $F : [0,1]^n \to [0,1]$ is called a pre-aggregation function if it satisfies boundary conditions $F(0, \ldots, 0) = 0$, $F(1, \ldots, 1) = 1$ and there exists a real vector $\overrightarrow{r} = (r_1, \ldots, r_n) \in [0,1]^n$, $\overrightarrow{r} \neq \overrightarrow{0}$ such that F is $\overrightarrow{r}$-increasing, i.e. for all $(x_1, \ldots, x_n) \in [0,1]^n$ and for all $c > 0$ such that $(x_1 + cr_1, \ldots, x_n + cr_n) \in [0,1]^n$ it holds

$$F(x_1 + cr_1, \ldots, x_n + cr_n) \geqslant F(x_1, \ldots, x_n).$$

Note that directional monotonicity (monotonicity along some vector) is a generalization of weak monotonicity (increasingness) introduced in [34], where the vector $\overrightarrow{r} = (1, \ldots, 1) \in [0, 1]^n$. Examples of proper pre-aggregation functions are: the mode (defined as the function that gives back the value which appears most times in the considered n-tuple, or the smallest of the values that appears most times, in case there is more than one such a value) or functions $F(x, y) = x - (\max(0, x - y))^2$, $F(x, y) = x \cdot |2y - 1|$, Lehmer mean $F(x, y) = \frac{x^2+y^2}{x+y}$ (weak monotonicity of Lehmer means was discussed in detail in [19, 20]). Other examples of pre-aggregation functions one may find in [33]. The reason to introduce pre-aggregation functions was the fact that some applications require non-monotone operations such as for example classification or image processing (cf. [33]). Such is the case when data may be corrupted by noise, or where we permit inputs to be classified as outliers and thus discounted in the average.

An overview of diverse types of means and aggregation functions, both the ones mentioned in this chapter and other classes, one may find in [20]. Moreover, recently other types of monotonicity and as a result new types of aggregation functions, namely *directionally ordered monotone aggregation functions*, were introduced. These type of functions allow monotonicity along different directions in different points. In particular, these functions take into account the ordinal size of the coordinates of the inputs in order to fuse them (cf. [35]). In the recent paper [36] a new notion of monotonicity, i.e. *strengthened ordered directional monotonicity* was proposed. This generalization of monotonicity is based on directional monotonicity and ordered directional monotonicity. Additionally, the family of *linear fusion functions* and the family of *ordered linear fusion functions*, as new types of operations on [0, 1] were introduced.

2.1.3 Dominance Between Aggregation Functions

Here, the notion of dominance is recalled. This is an interesting notion from mathematical point of view. This relation had been introduced for triangle functions in the framework of probabilistic metric spaces [37] and then for partially ordered sets [38]. It plays an important role in many fields of mathematics (cf. [39]) and it occurs as a useful toll for example in the preservation of various properties, most of them expressed by some inequalities, during aggregation processes in flexible querying, preference modelling and computer-assisted assessment (cf. [40]). These applications initiated the study of the dominance relation in the broader context of aggregation functions and even more generally for operations (cf. [41]). In this monograph the notion of dominance is used in Chap. 3.

Definition 2.14 (*cf.* [38]) Let $m, n \in \mathbb{N}$. A function $F\colon [0, 1]^m \to [0, 1]$ dominates function $G\colon [0, 1]^n \to [0, 1]$ ($F \gg G$) if for an arbitrary matrix $[x_{ik}] \in [0, 1]^{m\times n}$ the following inequality holds

$$F(G(x_{11},\ldots,x_{1n}),\ldots,G(x_{m1},\ldots,x_{mn})) \geqslant G(F(x_{11},\ldots,x_{m1}),\ldots,F(x_{1n},\ldots,x_{mn})). \tag{2.31}$$

If we replace in (2.31) the inequality $\geqslant$ with the equality $=$, then we get the commuting property [42]. Special case of the commuting property, when F equals G, is bisymmetry which has been already mentioned (cf. (2.25)). Here only a few examples of dominance between functions are presented. Other examples on dominance and comprehensive study of literature on this subject are gathered for example in [43].

Example 2.9 (*cf.* [40, 44]) A weighted geometric mean dominates t-norm T_P. A weighted arithmetic mean dominates t-norm T_L. The aggregation function

$$A(x_1,\ldots,x_n) = \frac{p}{n}\sum_{k=1}^{n} x_k + (1-p)\min_{1\leqslant k\leqslant n} x_k$$

dominates T_L, where $p \in (0,1)$. Let us consider projections P_k. Then $F \gg P_k$ and $P_k \gg F$ for any function $F : [0,1]^n \to [0,1]$.

Corollary 2.1 ([43]) *Minimum dominates any fuzzy conjunction. A weighted minimum (cf. [45])*

$$A(x_1,\ldots,x_n) = \min_{1\leqslant k\leqslant n} \max(1-w_k, x_k), \quad \max_{1\leqslant k\leqslant n} w_k = 1,$$

dominates any t-seminorm.

There exists a characterization theorem of increasing functions which dominate minimum.

Theorem 2.5 (cf. [40, Proposition 5.1]) *An increasing function $F\colon [0,1]^n \to [0,1]$ dominates minimum if and only if for each $x_1,\ldots,x_n \in [0,1]$*

$$F(x_1,\ldots,x_n) = \min(f_1(x_1),\ldots,f_n(x_n)),$$

where $f_k\colon [0,1] \to [0,1]$ is increasing with $k = 1,\ldots,n$.

Example 2.10 Examples of functions which dominate minimum are:

- $F = \min$, if $f_k(x) = x$, $k = 1,\ldots,n$,
- $F = P_k$ - projections, if for a certain $k \in \{1,\ldots,n\}$, function $f_k(x) = x$ and $f_i(x) = 1$ for $i \neq k$,
- F - a weighted minimum (cf. Corollary 2.1), if $f_k(x) = \max(1-w_k, x)$, $w_k \in [0,1]$, $k = 1,\ldots,n$, $\max_{1\leqslant k\leqslant n} w_k = 1$.

Dually, we get characterization theorem of increasing functions which are dominated by maximum.

Theorem 2.6 ([43]) *An increasing function* $F\colon [0,1]^n \to [0,1]$ *is dominated by maximum if and only if for each* $x_1, \ldots, x_n \in [0,1]$

$$F(x_1, \ldots, x_n) = \max(f_1(x_1), \ldots, f_n(x_n)),$$

where $f_k\colon [0,1] \to [0,1]$ *is increasing with* $k = 1, \ldots, n$.

Example 2.11 Examples of functions which are dominated by maximum are:

- $F = \max$, if $f_k(x) = x$, $k = 1, \ldots, n$,
- $F = P_k$ - projections, if for a certain $k \in \{1, \ldots, n\}$, function $f_k(x) = x$ and $f_i(x) = 1$ for $i \neq k$,
- F - a weighted maximum (cf. [45]), if $f_k(x) = \min(w_k, x)$, $w_k \in [0,1]$, $k = 1, \ldots, n$, $\max\limits_{1 \leqslant k \leqslant n} w_k = 1$, where

$$A(x_1, \ldots, x_n) = \max_{1 \leqslant k \leqslant n} \min(w_k, x_k), \quad \max_{1 \leqslant k \leqslant n} w_k = 1.$$

Aggregation functions proved to be effective tool in many application areas. For example, in information retrieval there are used diverse aggregation strategies [46]. Continually, there are introduced new classes of aggregation functions with good performance in applications, for example in fuzzy rule-based classification systems [47, 48]. Detailed information about aggregation functions are given in monographs devoted to this topic, for example [20, 49–51]. In this book the extensions of aggregation functions will be used, namely interval-valued aggregation functions which are recently developed by several authors and successfully applied (cf. [52, 53]). The notion of an aggregation function for interval-valued fuzzy settings is studied in the next section.

2.2 Classes of Aggregation Functions for Interval-Valued Fuzzy Settings

In this section there are presented some important classes of aggregation functions defined for interval-valued fuzzy settings. They will be also called *aggregation operators* which will be presented along with their properties. In interval-valued fuzzy settings we may use diverse representations and diverse orders while defining the concept of aggregation operator. The following notions are used throughout the monograph when regarding the aggregation operators and their formulas

$$\mathbf{x}_1 = [\underline{x}_1, \overline{x}_1], \mathbf{x}_2 = [\underline{x}_2, \overline{x}_2], \ldots, \mathbf{x}_n = [\underline{x}_n, \overline{x}_n].$$

2.2.1 Interval-Valued Aggregation Functions with Respect to the Classical Order

We start by recalling the notion of an interval-valued aggregation function with respect to the classical partial order $\preceq$ (cf. (1.8)). It will be also simply called an *interval-valued aggregation function* or an *aggregation function on* L^I. These are the most often used aggregation operators when regarding interval-valued fuzzy settings.

Definition 2.15 (*cf.* [54]) An operation $\mathscr{A} : (L^I)^n \to L^I$ is called an interval-valued aggregation function (or aggregation function on L^I), if it is increasing, i.e.

$$\forall_{\mathbf{x}_i, \mathbf{y}_i \in L^I} \; \mathbf{x}_i \preceq \mathbf{y}_i \Rightarrow \mathscr{A}(\mathbf{x}_1, \ldots, \mathbf{x}_n) \preceq \mathscr{A}(\mathbf{y}_1, \ldots, \mathbf{y}_n) \tag{2.32}$$

and it fulfils the conditions $\mathscr{A}(\underbrace{\mathbf{0}, \ldots, \mathbf{0}}_{n\times}) = \mathbf{0}, \quad \mathscr{A}(\underbrace{\mathbf{1}, \ldots, \mathbf{1}}_{n\times}) = \mathbf{1}.$

Among interval-valued aggregation functions we may consider the so called *representable* aggregation functions.

Definition 2.16 (*cf.* [55]) $\mathscr{A} : (L^I)^n \to L^I$ is said to be a representable aggregation function on L^I if there exist two aggregation functions $A_1, A_2 : [0, 1]^n \to [0, 1]$, $A_1 \leqslant A_2$ such that, for every $\mathbf{x}_1 = [\underline{x}_1, \overline{x}_1]$, $\mathbf{x}_2 = [\underline{x}_2, \overline{x}_2]$, ..., $\mathbf{x}_n = [\underline{x}_n, \overline{x}_n] \in L^I$ it holds that

$$\mathscr{A}(\mathbf{x}_1, \mathbf{x}_2, \ldots, \mathbf{x}_n) = [A_1(\underline{x}_1, \underline{x}_2, \ldots, \underline{x}_n), A_2(\overline{x}_1, \overline{x}_2, \ldots, \overline{x}_n)].$$

Functions A_1, A_2 will be called *components* (or *component functions*) of $\mathscr{A}$.

Lattice operations connected with the partial order $\preceq$ define representable aggregation functions on L^I, with $A_1 = A_2 = \min$ and $A_1 = A_2 = \max$. Moreover, many other examples of aggregation functions on L^I may be considered.

Example 2.12 Let $\mathbf{x}_1 = [\underline{x}_1, \overline{x}_1]$, $\mathbf{x}_2 = [\underline{x}_2, \overline{x}_2]$, ..., $\mathbf{x}_n = [\underline{x}_n, \overline{x}_n] \in L^I$. Thus one has:

- the representable product

$$\mathscr{A}([\underline{x}_1, \overline{x}_1], \ldots, [\underline{x}_n, \overline{x}_n]) = [\underline{x}_1 \cdot \ldots \cdot \underline{x}_n, \overline{x}_1 \cdot \ldots \cdot \overline{x}_n],$$

- the representable geometric mean

$$\mathscr{A}([\underline{x}_1, \overline{x}_1], \ldots, [\underline{x}_n, \overline{x}_n]) = [\sqrt{\underline{x}_1 \cdot \ldots \cdot \underline{x}_n}, \sqrt{\overline{x}_1 \cdot \ldots \cdot \overline{x}_n}],$$

- the representable product-mean

$$\mathscr{A}([\underline{x}_1, \overline{x}_1], \ldots, [\underline{x}_n, \overline{x}_n]) = \left[\underline{x}_1 \cdot \ldots \cdot \underline{x}_n, \frac{\overline{x}_1 + \cdots + \overline{x}_n}{n}\right].$$

In the literature one may find the notion of *decomposable* operations.

Definition 2.17 ([56]) An operation $\mathscr{F} : (L^I)^n \to L^I$ is called decomposable if there exist functions $F_1, F_2 : [0,1]^n \to [0,1]$ such that for any $\mathbf{x}_1 = [\underline{x}_1, \overline{x}_1]$, $\mathbf{x}_2 = [\underline{x}_2, \overline{x}_2]$, ..., $\mathbf{x}_n = [\underline{x}_n, \overline{x}_n] \in L^I$ it holds

$$\mathscr{F}(\mathbf{x}_1, \mathbf{x}_2, \ldots, \mathbf{x}_n) = [F_1(\underline{x}_1, \underline{x}_2, \ldots, \underline{x}_n), F_2(\overline{x}_1, \overline{x}_2, \ldots, \overline{x}_n)].$$

Functions F_1, F_2 will be called *components* (or *component functions*) of $\mathscr{F}$.

Certainly, definition of decomposable functions makes sense only if $F_1 \leqslant F_2$. Note that a representable aggregation function is a particular instance of a decomposable operation. Moreover, an aggregation function is decomposable if and only if it is representable, as the next result shows.

Theorem 2.7 (cf. [57]) *An operation $\mathscr{A} : (L^I)^2 \to L^I$ is a decomposable aggregation function if and only if there exist aggregation functions $A_1, A_2 : [0,1]^n \to [0,1]$ such that for any $\boldsymbol{x}_1 = [\underline{x}_1, \overline{x}_1]$, $\boldsymbol{x}_2 = [\underline{x}_2, \overline{x}_2]$,..., $\boldsymbol{x}_n = [\underline{x}_n, \overline{x}_n] \in L^I$ and $A_1 \leqslant A_2$ it holds*

$$\mathscr{A}(\boldsymbol{x}_1, \boldsymbol{x}_2, \ldots, \boldsymbol{x}_n) = [A_1(\underline{x}_1, \underline{x}_2, \ldots, \underline{x}_n), A_2(\overline{x}_1, \overline{x}_2, \ldots, \overline{x}_n)].$$

Remark 2.2 Since decomposability for aggregation functions is equivalent to representability, in the sequel the notion of decomposability will be used when referring to general notion of operation and also when referring to aggregation operators. The notion of representability will be used when referring to the notion of aggregation functions (cf. Definitions 2.15, 2.16).

Below, we present other examples of representable interval-valued aggregation functions which are built using Theorem 2.7.

Example 2.13 (*cf.* [58]) Examples of representable interval-valued aggregation functions are the following:

$$\mathscr{A}(\mathbf{x}_1, \mathbf{x}_2, \ldots, \mathbf{x}_n) = \left[\frac{\underline{x}_1 + \underline{x}_2 + \cdots + \underline{x}_n}{n}, \frac{\overline{x}_1 + \overline{x}_2 + \cdots + \overline{x}_n}{n}\right], \tag{2.33}$$

$$\mathscr{A}(\mathbf{x}_1, \mathbf{x}_2, \ldots, \mathbf{x}_n) = \left[\sqrt{\frac{\underline{x}_1^2 + \underline{x}_2^2 + \cdots + \underline{x}_n^2}{n}}, \sqrt[3]{\frac{\overline{x}_1^3 + \overline{x}_2^3 + \cdots + \overline{x}_n^3}{n}}\right], \tag{2.34}$$

$$\mathscr{A}(\mathbf{x}_1, \mathbf{x}_2, \ldots, \mathbf{x}_n) = \left[\sqrt[3]{\frac{\underline{x}_1^3 + \underline{x}_2^3 + \cdots + \underline{x}_n^3}{n}}, \sqrt[4]{\frac{\overline{x}_1^4 + \overline{x}_2^4 + \cdots + \overline{x}_n^4}{n}}\right]. \tag{2.35}$$

where (2.33) is called the *interval-valued arithmetic mean* and (2.34), (2.35) are constructed with the use of the adequate power means (cf. Example 2.3). For these family of means we have the monotonicity dependence of the form (2.14). According to this dependence, the lower and upper bound of the operator (2.34) were obtained for $r = 2$ and $s = 3$, respectively. Similarly, the lower and upper bound of the operator (2.35) were obtained for $r = 3$ and $s = 4$, respectively. Analogously, using (2.13) and (2.14), parameterized families of representable aggregation functions may be constructed. This is presented in the formula (2.36).

Example 2.14 Let $0 < w_k < 1$ for $k = 1, \dots, n$, $\sum_{k=1}^{n} w_k = 1$, $\underline{a} + \underline{b} + \underline{c} = 1$, $\underline{a}, \underline{b}, \underline{c} \in (0, 1)$, $\overline{a} + \overline{b} + \overline{c} = 1$, $\overline{a}, \overline{b}, \overline{c} \in (0, 1)$, $\underline{a} \leqslant \overline{a}$, $\underline{b} \leqslant \overline{b}$, $\underline{c} \leqslant \overline{c}$, and $\left(\sum_{k=1}^{n} \frac{w_k}{\underline{x}_k}\right)^{-1} = 0$ if there exits at least one element $\underline{x}_k = 0$. Thus the following operator is a representable interval-valued aggregation function

$$\mathscr{A}(\mathbf{x}_1, \mathbf{x}_2, \dots, \mathbf{x}_n) =$$

$$\left[\underline{a} \cdot \prod_{k=1}^{n} \underline{x}_k^{w_k} + \underline{b} \cdot \sum_{k=1}^{n} w_k \underline{x}_k + \underline{c} \cdot \left(\sum_{k=1}^{n} \frac{w_k}{\underline{x}_k}\right)^{-1}, \overline{a} \cdot \sum_{k=1}^{n} w_k \overline{x}_k + \overline{b} \cdot \sqrt{\sum_{k=1}^{n} w_k \overline{x}_k^2} + \overline{c} \cdot \sqrt[3]{\sum_{k=1}^{n} w_k \overline{x}_k^3}\right]. \tag{2.36}$$

Let us notice that the interval-valued arithmetic mean, given by (2.33), could be obtained from the formula (2.36) for $\underline{a} = \underline{c} = \overline{b} = \overline{c} = 0$, $\underline{b} = \overline{a} = 1$ and equal weights $w_k = \frac{1}{n}$ for $k = 1, \dots, n$. A notable examples of binary interval-valued aggregation functions are triangular norms and conorms.

Definition 2.18 (*cf.* [59]) A representable triangular norm $\mathscr{T}$ on L^I is an increasing (cf. (2.32)), symmetric and associative operation $\mathscr{T} : (L^I)^2 \to L^I$ with a neutral element $\mathbf{1}$. A representable triangular conorm $\mathscr{S}$ on L^I is an increasing, symmetric and associative operation $\mathscr{S} : (L^I)^2 \to L^I$ with a neutral element $\mathbf{0}$.

Properties of symmetry, associativity, neutral element are defined analogously to the case of adequate properties of operations on [0, 1] (cf. Sect. 2.5).

Let $\mathbf{x} = [\underline{x}, \overline{x}]$ and $\mathbf{y} = [\underline{y}, \overline{y}]$. The following are examples of representable triangular norms on L^I:

- $\mathscr{T}(\mathbf{x}, \mathbf{y}) = [\min(\underline{x}, \underline{y}), \min(\overline{x}, \overline{y})]$,
- $\mathscr{T}(\mathbf{x}, \mathbf{y}) = [\max(0, \underline{x} + \underline{y} - 1), \min(\overline{x}, \overline{y})]$,
- $\mathscr{T}(\mathbf{x}, \mathbf{y}) = [\max(0, \underline{x} + \underline{y} - 1), \max(0, \overline{x} + \overline{y} - 1)]$.

Below there are some examples of representable triangular conorms on L^I:

- $\mathscr{T}(\mathbf{x}, \mathbf{y}) = [\max(\underline{x}, \underline{y}), \max(\overline{x}, \overline{y})]$,
- $\mathscr{T}(\mathbf{x}, \mathbf{y}) = [\max(\underline{x}, \underline{y}), \min(\overline{x} + \overline{y}, 1)]$,
- $\mathscr{T}(\mathbf{x}, \mathbf{y}) = [\min(\underline{x} + \underline{y}, 1), \min(\overline{x} + \overline{y}, 1)]$.

Representability is not the only possible way to build interval-valued aggregation functions. In the literature there are also considered *pseudomax representable* and *pseudomin representable aggregation functions*.

Definition 2.19 (*cf.* [57]) Let $\mathbf{x}_1 = [\underline{x}_1, \overline{x}_1]$, $\mathbf{x}_2 = [\underline{x}_2, \overline{x}_2]$,..., $\mathbf{x}_n = [\underline{x}_n, \overline{x}_n] \in L^I$ and let $A_1, A_2 : [0, 1]^n \to [0, 1]$, $A_1 \leqslant A_2$ be aggregation functions. The aggregation function $\mathscr{A}$ is called a pseudomax A_1A_2-representable and a pseudomin A_1A_2-representable, respectively if

$$\mathscr{A}(\mathbf{x}_1, \mathbf{x}_2, \ldots, \mathbf{x}_n) = \\ [A_1(\underline{x}_1, \underline{x}_2, \ldots, \underline{x}_n), \max(A_2(\underline{x}_1, \overline{x}_2, \ldots, \overline{x}_n), A_2(\overline{x}_1, \underline{x}_2, \overline{x}_3, \ldots, \overline{x}_n), \ldots, A_2(\overline{x}_1, \ldots, \overline{x}_{n-1}, \underline{x}_n))], \tag{2.37}$$

$$\mathscr{A}(\mathbf{x}_1, \mathbf{x}_2, \ldots, \mathbf{x}_n) = \\ [\min(A_1(\overline{x}_1, \underline{x}_2, \ldots, \underline{x}_n), A_1(\underline{x}_1, \overline{x}_2, \underline{x}_3, \ldots, \underline{x}_n), \ldots, A_1(\underline{x}_1, \ldots, \underline{x}_{n-1}, \overline{x}_n)), A_2(\overline{x}_1, \overline{x}_2, \ldots, \overline{x}_n)]. \tag{2.38}$$

Functions A_1, A_2 will be called components (or component functions) of $\mathscr{A}$.

Among pseudomax and pseudomin-representable interval-valued aggregation functions we also have pseudo-t-representable norms and pseudo-t-representable conorms (cf. [60]). Moreover, we have the following examples of construction methods for interval-valued aggregation functions.

Example 2.15 (*cf.* [57]) Let $\mathbf{x}_1 = [\underline{x}_1, \overline{x}_1]$, $\mathbf{x}_2 = [\underline{x}_2, \overline{x}_2]$,..., $\mathbf{x}_n = [\underline{x}_n, \overline{x}_n] \in L^I$, $A_i : [0, 1]^n \to [0, 1]$ be aggregation functions, $i \in \{1, \ldots, 4\}$, $A_1 \leqslant A_2$, $A_3 \leqslant A_4$. The following are aggregation functions on L^I:

$$\mathscr{A}(\mathbf{x}_1, \mathbf{x}_2, \ldots, \mathbf{x}_n) = [\min(A_1(\overline{x}_1, \underline{x}_2, \ldots, \underline{x}_n), A_1(\underline{x}_1, \overline{x}_2, \underline{x}_3, \ldots, \underline{x}_n), \ldots, A_1(\underline{x}_1, \ldots, \underline{x}_{n-1}, \overline{x}_n)), \\ \max(A_2(\underline{x}_1, \overline{x}_2, \ldots, \overline{x}_n), A_2(\overline{x}_1, \underline{x}_2, \overline{x}_3, \ldots, \overline{x}_n), \ldots, A_2(\overline{x}_1, \ldots, \overline{x}_{n-1}, \underline{x}_n))], \tag{2.39}$$

$$\mathscr{A}(\mathbf{x}_1, \mathbf{x}_2, \ldots, \mathbf{x}_n) = [A_3(A_1(\overline{x}_1, \underline{x}_2, \ldots, \underline{x}_n), A_1(\underline{x}_1, \overline{x}_2, \underline{x}_3, \ldots, \underline{x}_n), \ldots, A_1(\underline{x}_1, \ldots, \underline{x}_{n-1}, \overline{x}_n)), \\ A_4(A_2(\underline{x}_1, \overline{x}_2, \ldots, \overline{x}_n), A_2(\overline{x}_1, \underline{x}_2, \overline{x}_3, \ldots, \overline{x}_n), \ldots, A_2(\overline{x}_1, \ldots, \overline{x}_{n-1}, \underline{x}_n))]. \tag{2.40}$$

Aggregation function (2.39) is a special case of (2.40).

In the next subsections other classes of aggregation operators for interval-valued fuzzy settings are presented. These are possible and necessary aggregation functions and interval-valued aggregation functions with respect to linear orders.

2.2.2 Pos-Aggregation Functions and Nec-Aggregation Functions

New classes of aggregation operators may be obtained by replacing in the monotonicity condition (2.32) the partial order $\preceq$ with the comparability relations $\preceq_\pi$ or

$\preceq_\nu$. Note that the aggregation operators obtained in such way are not special cases of aggregation operators on lattices (well described in the literature), since relations $\preceq_\pi$ and $\preceq_\nu$ are not partial orders (cf. Proposition 1.1).

Definition 2.20 ([61]) An operation $\mathscr{A} : (L^I)^n \to L^I$ is called a possible aggregation function (we will write for short pos-aggregation function) if

$$\underset{\mathbf{x}_i,\mathbf{y}_i\in L^I}{\forall} \mathbf{x}_i \preceq_\pi \mathbf{y}_i \Rightarrow \mathscr{A}(\mathbf{x}_1,\ldots,\mathbf{x}_n) \preceq_\pi \mathscr{A}(\mathbf{y}_1,\ldots,\mathbf{y}_n) \tag{2.41}$$

and $\mathscr{A}(\underbrace{\mathbf{0},\ldots,\mathbf{0}}_{n\times}) = \mathbf{0}, \quad \mathscr{A}(\underbrace{\mathbf{1},\ldots,\mathbf{1}}_{n\times}) = \mathbf{1}.$

Definition 2.21 ([61]) An operation $\mathscr{A} : (L^I)^n \to L^I$ is called a necessary aggregation function (we will write for short nec-aggregation function) if

$$\underset{\mathbf{x}_i,\mathbf{y}_i\in L^I}{\forall} \mathbf{x}_i \preceq_\nu \mathbf{y}_i \Rightarrow \mathscr{A}(\mathbf{x}_1,\ldots,\mathbf{x}_n) \preceq_\nu \mathscr{A}(\mathbf{y}_1,\ldots,\mathbf{y}_n) \tag{2.42}$$

and $\mathscr{A}(\underbrace{\mathbf{0},\ldots,\mathbf{0}}_{n\times}) = \mathbf{0}, \quad \mathscr{A}(\underbrace{\mathbf{1},\ldots,\mathbf{1}}_{n\times}) = \mathbf{1}.$

The family of all pos-aggregation functions will be denoted by $\mathscr{A}_\pi$ and the family of all nec-aggregation functions will be denoted by $\mathscr{A}_\nu$.

Example 2.16 The following operations are examples of pos-aggregation functions:

$$\mathscr{A}(\mathbf{x}_1,\mathbf{x}_2,\ldots,\mathbf{x}_n) = \left[\frac{\underline{x}_1+\underline{x}_2+\cdots+\underline{x}_n}{n}, \frac{\overline{x}_1^2+\overline{x}_2^2+\cdots+\overline{x}_n^2}{\overline{x}_1+\overline{x}_2+\cdots+\overline{x}_n}\right], \tag{2.43}$$

$$\mathscr{A}(\mathbf{x}_1,\mathbf{x}_2,\ldots,\mathbf{x}_n) = \left[\frac{\underline{x}_1+\underline{x}_2+\cdots+\underline{x}_n}{n}, \frac{\overline{x}_1^3+\overline{x}_2^3+\cdots+\overline{x}_n^3}{\overline{x}_1^2+\overline{x}_2^2+\cdots+\overline{x}_n^2}\right]. \tag{2.44}$$

Pos-aggregation functions (2.43) and (2.44) are defined with the convention $\frac{0}{0} = 0$, where $\frac{0}{0}$ may occur if it holds $\mathbf{x}_1 = \mathbf{x}_2 = \cdots = \mathbf{x}_n = [0,0]$, as a result if $\overline{x}_1 = \overline{x}_2 = \cdots = \overline{x}_n = 0$. Operators (2.43) and (2.44) are not interval-valued aggregation functions since the upper bounds of the intervals are not monotonic according to (2.22). The upper bounds of these aggregation operators are built with the use of *Lehmer means* presented in Example 2.4 (in [19, 20], p. 261 monotonicity of Lehmer means was discussed in detail). Moreover, possible aggregation functions (2.43) and (2.44) are built using dependence (2.19).

More sophisticated examples, i.e. parameterized families of pos-aggregation functions may be obtained in the following way.

Example 2.17 Let $r \in \mathbb{R}$, $r > 1$, $\underline{a} + \underline{b} + \underline{c} = 1$, $\underline{a}, \underline{b}, \underline{c} \in (0, 1)$, $\overline{a} + \overline{b} + \overline{c} = 1$, $\overline{a}, \overline{b}, \overline{c} \in (0, 1)$, $\underline{a} \leqslant \overline{a}, \underline{b} \leqslant \overline{b}, \underline{c} \leqslant \overline{c}$, and it holds $\left(\frac{1}{n}\sum_{k=1}^{n}\frac{1}{\underline{x}_k}\right)^{-1} = 0$, if there exists at least one element $\underline{x}_k = 0$. Thus the following operation is a pos-aggregation function

$$\mathscr{A}(\mathbf{x}_1, \mathbf{x}_2, \ldots, \mathbf{x}_n) =$$
$$\left[\underline{a} \cdot \sqrt[n]{\prod_{k=1}^{n} \underline{x}_k} + \underline{b} \cdot \frac{1}{n}\sum_{k=1}^{n} \underline{x}_k + \underline{c} \cdot \left(\frac{1}{n}\sum_{k=1}^{n}\frac{1}{\underline{x}_k}\right)^{-1}, \overline{a} \cdot \frac{\sum_{k=1}^{n} \overline{x}_k^2}{\sum_{k=1}^{n} \overline{x}_k} + \overline{b}\frac{\sum_{k=1}^{n} \overline{x}_k^3}{\sum_{k=1}^{n} \overline{x}_k^2} + \overline{c} \cdot \frac{\sum_{k=1}^{n} \overline{x}_k^r}{\sum_{k=1}^{n} \overline{x}_k^{r-1}}\right]. \tag{2.45}$$

Let us note that the upper bound of the interval in the formula (2.45) is constructed with the use of the Lehmer means (which fail the standard monotonicity of aggregation functions on [0, 1], cf. (2.22)). Thus $\mathscr{A}$ given by the formula (2.45) is not an interval-valued aggregation function but it is a pos-aggregation function. Moreover, the formula (2.45) was obtained using dependencies (2.18) and (2.19). Formulas (2.43) and (2.44) could be obtained from (2.45) for $\underline{a} = \underline{c} = 0, \underline{b} = 1$ and $\overline{a} = 1, \overline{b} = \overline{c} = 0$ (or $\overline{b} = 1, \overline{a} = \overline{c} = 0$).

Example 2.18 The following operation is a nec-aggregation function (note that this function is a pseudomax A_1A_2-representable aggregation function, where $A_1 = A_2$ is the arithmetic mean)

$$\mathscr{A}(\mathbf{x}_1, \mathbf{x}_2, \ldots, \mathbf{x}_n) =$$
$$\left[\frac{\underline{x}_1 + \underline{x}_2 + \cdots + \underline{x}_n}{n}, \max\left(\frac{\underline{x}_1 + \overline{x}_2 + \cdots + \overline{x}_n}{n}, \frac{\overline{x}_1 + \underline{x}_2 + \overline{x}_3 + \cdots + \overline{x}_n}{n}, \ldots, \frac{\overline{x}_1 + \cdots + \overline{x}_{n-1} + \underline{x}_n}{n}\right)\right]. \tag{2.46}$$

Examples of parameterized families of nec-aggregation functions are presented below (cf. (2.47) and (2.48)). They are obtained using Definition 2.19, Example 2.3 and Proposition 2.6.

Example 2.19 Let $r \in \mathbb{R}_0$, $0 < w_k < 1$ for $k = 1, \ldots, n$, $\sum_{k=1}^{n} w_k = 1$. Thus the following operator is a nec-aggregation function

$$\mathscr{A}(\mathbf{x}_1, \mathbf{x}_2, \ldots, \mathbf{x}_n) = \left[\frac{1}{r}\ln\left(\sum_{k=1}^{n} w_k e^{r\underline{x}_k}\right),\right.$$
$$\left.\frac{1}{r}\max\left(\ln\left(w_1 e^{r\underline{x}_1} + \sum_{k=2}^{n} w_k e^{r\overline{x}_k}\right), \ln\left(w_2 e^{r\underline{x}_2} + \sum_{k=1,k\neq 2}^{n} w_k e^{r\overline{x}_k}\right), \ldots, \ln\left(w_n e^{r\underline{x}_n} + \sum_{k=1}^{n-1} w_k e^{r\overline{x}_k}\right)\right)\right]. \tag{2.47}$$

The other parameterized family of nec-aggregation functions is given as follows.

Example 2.20 Let $r \in \mathbb{R}$, $r > 0$, $0 < w_k < 1$ for $k = 1, \ldots, n$, $\sum_{k=1}^{n} w_k = 1$. The following operator is a nec-aggregation function

$$\mathscr{A}(\mathbf{x}_1, \mathbf{x}_2, \ldots, \mathbf{x}_n) =$$
$$\left[\left(\sum_{k=1}^{n} w_k \underline{x}_k^r\right)^{\frac{1}{r}}, \max\left(\left(w_1\underline{x}_1^r + \sum_{k=2}^{n} w_k \overline{x}_k^r\right)^{\frac{1}{r}}, \left(w_2\underline{x}_2^r + \sum_{k=1, k\neq 2}^{n} w_k \overline{x}_k^r\right)^{\frac{1}{r}}, \ldots, \left(w_n\underline{x}_n^r + \sum_{k=1}^{n-1} w_k \overline{x}_k^r\right)^{\frac{1}{r}}\right)\right]. \tag{2.48}$$

More examples of nec-aggregation functions are provided in Sect. 2.3.

2.2.3 *Interval-Valued Aggregation Functions with Respect to Linear Orders*

Recently, interval-valued aggregation functions with respect to linear orders, especially admissible linear orders, were introduced. New types of aggregation operators are obtained by replacing in the monotonicity condition (2.32) the partial order $\preceq$ with the admissible linear orders.

Definition 2.22 ([62]) $\mathscr{A} : (L^I)^n \to L^I$ is called an aggregation function with respect to the admissible linear order $\leq_{L^I}$, if

$$\forall_{\mathbf{x}_i, \mathbf{y}_i \in L^I} \ \mathbf{x}_i \leq_{L^I} \mathbf{y}_i \Rightarrow \mathscr{A}(\mathbf{x}_1, \ldots, \mathbf{x}_n) \leq_{L^I} \mathscr{A}(\mathbf{y}_1, \ldots, \mathbf{y}_n)$$

and $\mathscr{A}(\underbrace{\mathbf{0}, \ldots, \mathbf{0}}_{n\times}) = \mathbf{0}, \quad \mathscr{A}(\underbrace{\mathbf{1}, \ldots, \mathbf{1}}_{n\times}) = \mathbf{1}$.

Example 2.21 ([62]) Let $p > 0, n = 2$. The following operations are interval-valued aggregation functions with respect to the given linear orders:

- $\mathscr{A}([\underline{x}, \overline{x}], [\underline{y}, \overline{y}]) = [(\underline{x}\underline{y})^p, (\overline{x}\overline{y})^p]$ with respect to $\leq_{Lex2}$,
- $\mathscr{A}([\underline{x}, \overline{x}], [\underline{y}, \overline{y}]) = [\frac{\underline{x}\overline{y}+\overline{x}\underline{y}}{2}, \frac{\underline{x}\underline{y}+\overline{x}\overline{y}}{2}]$, with respect to $\leq_{XY}$
- $\mathscr{A}([\underline{x}, \overline{x}], [\underline{y}, \overline{y}]) = [\alpha\underline{x} + (1-\alpha)\underline{y}, \alpha\overline{x} + (1-\alpha)\overline{y}]$ with respect to $\leq_{XY}$, $\leq_{Lex1}$ and $\leq_{Lex2}$, where $\alpha \in [0, 1]$.

Below we present sufficient conditions for a decomposable aggregation operator to be linear with respect to the orders $\leq_{Lex1}$ or $\leq_{Lex2}$.

Proposition 2.2 *Let* $\mathscr{A} : (L^I)^n \to L^I$ *be a decomposable operator with component functions* A_1, A_2. *If the component function* A_1 *is a strictly increasing aggregation function on* $[0, 1]$ *and the component function* A_2 *is an aggregation function on* $[0, 1]$, *then* $\mathscr{A}$ *is an interval-valued aggregation function with respect to the linear order* $\leq_{Lex1}$.

Proof For simplicity of notions we present the proof for a binary decomposable operator $\mathscr{A}(\mathbf{x}, \mathbf{y}) = [A_1(\underline{x}, \underline{y}), A_2(\overline{x}, \overline{y})]$. Let $\mathscr{A}$ be a decomposable operator with component functions A_1, A_2 and A_1 be a strictly increasing aggregation function, A_2

be an aggregation function. We will show that for $\mathbf{x}, \mathbf{y}, \mathbf{u}, \mathbf{w} \in L^I$ such that $\mathbf{x} \leq_{Lex1} \mathbf{y}$ and $\mathbf{u} \leq_{Lex1} \mathbf{w}$ we obtain $\mathscr{A}(\mathbf{x}, \mathbf{u}) \leq_{Lex1} \mathscr{A}(\mathbf{y}, \mathbf{w})$. We have the following cases:
(1) $\underline{x} < \underline{y}$ and $\underline{u} < \underline{w}$ imply $A_1(\underline{x}, \underline{u}) < A_1(\underline{y}, \underline{w})$ (since A_1 is strictly increasing, cf. (2.30));
(2) $\underline{x} < \underline{y}, \underline{u} = \underline{w}$ and $\overline{u} \leqslant \overline{w}$ imply $A_1(\underline{x}, \underline{u}) < A_1(\underline{y}, \underline{w})$ (since A_1 is strictly increasing);
(3) $\underline{x} = \underline{y}, \overline{x} \leqslant \overline{y}$ and $\underline{u} < \underline{w}$ imply $A_1(\underline{x}, \underline{u}) < A_1(\underline{y}, \underline{w})$ (since A_1 is strictly increasing);
(4) $\underline{x} = \underline{y}, \overline{x} \leqslant \overline{y}, \underline{u} = \underline{w}$ and $\overline{u} \leqslant \overline{w}$ - under given assumptions $A_1(\underline{x}, \underline{u}) = A_1(\underline{y}, \underline{w})$, so since A_2 is increasing we get $A_2(\overline{x}, \overline{u}) \leqslant A_2(\overline{y}, \overline{w})$. As a result the monotonicity condition for $\mathscr{A}$ was proved. Boundary conditions for $\mathscr{A}$ are obtained by assumptions of A_1 and A_2, which are aggregation functions.

Proposition 2.3 *Let $\mathscr{A} : (L^I)^n \to L^I$ be a decomposable operator with component functions A_1 and A_2. If the component function A_1 is an aggregation function on $[0, 1]$ and the component function A_2 is a strictly increasing aggregation function on $[0, 1]$, then $\mathscr{A}$ is an interval-valued aggregation function with respect to the linear order $\leq_{Lex2}$.*

Proof The justification is analogous to the one presented in Proposition 2.2 for the linear order $\leq_{Lex1}$.

In the following example we will discuss the importance of the assumption of strict monotonicity for component aggregation functions (cf. Propositions 2.2 and 2.3).

Example 2.22 Let us consider the aggregation function $\mathscr{A}(\mathbf{x}, \mathbf{y}) = [\underline{x}\,\underline{y}, \overline{x}\,\overline{y}]$, where $A_1 = A_2 = T_P$ which is a product t-norm and it is a strictly increasing aggregation function on $(0, 1]$ (but not on $[0, 1]$, cf. Example 2.7). $\mathscr{A}$ is an interval-valued aggregation function with respect to the order $\leq_{Lex2}$ (cf. the first item in Example 2.21 for $p = 1$). This means that condition given in Proposition 2.3 for $\mathscr{A}$ to be an interval-valued aggregation function with respect to the order $\leq_{Lex2}$ is not necessary.

However, $\mathscr{A}(\mathbf{x}, \mathbf{y}) = [\underline{x}\,\underline{y}, \overline{x}\,\overline{y}]$ is not an interval-valued aggregation function with respect to the order $\leq_{Lex1}$. Let us consider $\mathbf{x} = [0.4, 0.8]$, $\mathbf{y} = [0.5, 0.6]$ and $\mathbf{u} = \mathbf{w} = [0, 1]$. Thus $\mathbf{x} \leq_{Lex1} \mathbf{y}$ and $\mathbf{u} \leq_{Lex1} \mathbf{w}$ but $\mathscr{A}(\mathbf{y}, \mathbf{w}) = [0, 0.6] \leq_{Lex1} [0, 0.8] = \mathscr{A}(\mathbf{x}, \mathbf{u})$.

By Propositions 2.2 and 2.3 we may obtain many examples of interval-valued aggregation functions with respect to the linear orders $\leq_{Lex1}$ or $\leq_{Lex2}$. Since linear-means are strictly increasing on $[0, 1]$ (cf. (2.11), Examples 2.3, and 2.7), by monotonicity conditions (2.13) and (2.14) we may obtain the following parameterized family of interval-valued aggregation functions with respect to the linear orders $\leq_{Lex1}$ or $\leq_{Lex2}$.

Example 2.23 Let $0 < r < s, r, s \in \mathbb{R}$. The following operator is an interval-valued aggregation function with respect to the linear order $\leq_{Lex1}$ or $\leq_{Lex2}$, where

$$\mathscr{A}(\mathbf{x}_1,\ldots,\mathbf{x}_n)=\left[\sqrt[r]{\sum_{k=1}^{n} w_k\underline{x}_k^r},\ \sqrt[s]{\sum_{k=1}^{n} w_k\overline{x}_k^s}\right].$$

As a useful example of an aggregation function with respect to an admissible linear order we may put IVOWA operator.

Definition 2.23 ([63]) Let $\leq_{L^I}$ be an admissible order on L^I, and let be given a vector $w=(w_1,\ldots,w_n)\in[0,1]^n$, with $w_1+\cdots+w_n=1$. The Interval-Valued OWA operator (IVOWA) associated with $\leq_{L^I}$ and w is a mapping $IVOWA_{\leq_{L^I},w}:(L^I)^n\to L^I$, given by

$$IVOWA_{\leq_{L^I},w}([\underline{x}_1,\overline{x}_1],\ldots,[\underline{x}_n,\overline{x}_n])=\sum_{i=1}^{n} w_i\cdot[\underline{x}_{(i)},\overline{x}_{(i)}],$$

where $[\underline{x}_{(i)},\overline{x}_{(i)}]$, $i=1,\ldots,n$, denotes the ith greatest of the inputs with respect to the order $\leq_{L^I}$ and $w\cdot[\underline{x},\overline{x}]=[w\underline{x},w\overline{x}]$, $[\underline{x}_1,\overline{x}_1]+[\underline{x}_2,\overline{x}_2]=[\underline{x}_1+\underline{x}_2,\overline{x}_1+\overline{x}_2]$.

Let us note that $IVOWA_{\leq_{L^I},w}$ is not an aggregation function with respect to the partial order $\preceq$.

Example 2.24 ([63]) Let us consider the admissible linear order $\leq_{XY}$, the weighting vector $w=(0.8,0.2)$ and intervals $\mathbf{x}=[0.5,0.5]$, $\mathbf{y}=[0.1,1]$ and $\mathbf{z}=[0.6,0.6]$. Thus $\mathbf{x}\leq_{XY}\mathbf{y}\leq_{XY}\mathbf{z}$ and we have $IVOWA_{\leq_{XY},w}(\mathbf{x},\mathbf{y})=0.8\cdot[0.1,1]+0.2\cdot[0.5,0.5]=[0.18,0.9]$, $IVOWA_{\leq_{XY},w}(\mathbf{z},\mathbf{y})=0.8\cdot[0.6,0.6]+0.2\cdot[0.1,1]=[0.5,0.68]$. We see that although $\mathbf{x}\preceq\mathbf{z}$, i.e. we have increased the first input of intervals with respect to the order $\preceq$, the obtained values of $IVOWA_{\leq_{XY},w}(\mathbf{x},\mathbf{y})$ and $IVOWA_{\leq_{XY},w}(\mathbf{z},\mathbf{y})$ are not comparable with respect to the order $\preceq$. It means that $IVOWA_{\leq_{XY},w}$ is not an aggregation function with respect to the partial order $\preceq$.

Moreover, $IVOWA_{\leq_{L^I},w}$ may be an aggregation function only with respect to some types of linear orders.

Proposition 2.4 ([63]) *Let $\leq_{\alpha\beta}$ be an admissible linear order on L^I (cf. Example 1.5). Thus the operator $IVOWA_{\leq_{\alpha\beta},w}$ is an aggregation function with respect to the order $\leq_{\alpha\beta}$.*

IVOWA operator may be also build with the use of an interval-valued overlap operator [64], where the latter is a conjunction operator defined for interval-valued settings in [65]. Moreover, in [65] there is also introduced the concept of a grouping function for interval-valued fuzzy settings (overlap and grouping functions are subclasses of conjunctions and disjunctions on [0, 1], respectively [66, 67]). In that paper there are also considered overlap and grouping operators for interval-valued fuzzy settings fulfilling some types of monotonicity. For example, the following interval-valued function $\mathscr{F}:(L^I)^2\to L^I$ (given in [68]) is increasing with respect to an inclusion order $\preceq_2$ (it is also an interval-valued overlap function [65]), where

$$\mathscr{F}(\mathbf{x}, \mathbf{y}) = \begin{cases} [\underline{x} + \underline{y} - 1, \overline{xy}] & \text{if } \underline{x} + \underline{y} > 1 \\ [0, \overline{xy}] & \text{otherwise.} \end{cases}$$

In the next section we will consider all presented so far types of aggregation operators for interval-valued fuzzy settings with regard to dependencies between them.

2.3 Dependencies Between Classes of Aggregation Functions in Interval-Valued Fuzzy Settings

This section presents dependencies between the considered in this book families of aggregation operators used in interval-valued fuzzy settings, i.e. pos- and nec-aggregation functions and aggregation functions with respect to $\preceq$ and $\leq_{L^I}$. For the simplicity of notions two-argument version of the considered aggregation operators will be used.

2.3.1 Interval-Valued Aggregation Functions Versus Pos-Aggregation Functions and Nec-Aggregation Functions

It is easy to check that the considered families of aggregation functions $\mathscr{A}_\pi$ and $\mathscr{A}_\nu$ have the same bounds, which are decomposable operators.

Proposition 2.5 ([61]) *Decomposable aggregation operators $\mathscr{A}_w, \mathscr{A}_s : (L^I)^2 \to L^I$ are respectively the lower and upper bounds of the families $\mathscr{A}_\pi$, $\mathscr{A}_\nu$ and the family of all interval-valued aggregation functions, where*

$$\mathscr{A}_w(\boldsymbol{x}, \boldsymbol{y}) = \begin{cases} [1, 1], & (x, y) = ([1, 1], [1, 1]) \\ [0, 0], & \textit{otherwise}, \end{cases}$$

$$\mathscr{A}_s(\boldsymbol{x}, \boldsymbol{y}) = \begin{cases} [0, 0], & (x, y) = ([0, 0], [0, 0]) \\ [1, 1], & \textit{otherwise}, \end{cases}$$

and $\mathscr{A}_w = [A_w, A_w]$, $\mathscr{A}_s = [A_s, A_s]$, A_w is the weakest aggregation function on $[0, 1]$ *and A_s is the strongest aggregation function on* $[0, 1]$ *(cf. Proposition 2.1).*

Implications given in Proposition 1.1 suggest that similar implications may be obtained for families of interval-valued aggregation functions, pos-aggregation functions and nec-aggregation functions. However, this is true only for decomposable operators (cf. Corollaries 2.2, 2.3). In other cases these connections are more complicated. We will use the notation $\mathscr{A}^*$ to gather aggregation operators on L^I of the type (2.37)–(2.40) and $\mathscr{D}$ to gather decomposable operators. We will recall some of the examples of operators and dependencies between them presented in [61]. This will be Examples 2.25–2.42.

Example 2.25 The following non-decomposable operation is an aggregation function on L^I (cf. Definition 2.15), where $\mathbf{x} = [\underline{x}, \overline{x}], \mathbf{y} = [\underline{y}, \overline{y}], \mathscr{A}(\mathbf{x}, \mathbf{y}) = [\underline{y}\frac{\underline{x}+\overline{x}}{2}, \frac{\underline{x}+\overline{x}}{2}]$, but $\mathscr{A} \notin \mathscr{A}^*$ and it also does not belong to other considered here families of aggregation functions, i.e. $\mathscr{A} \notin \mathscr{A}_\pi, \mathscr{A} \notin \mathscr{A}_\nu$.

Example 2.26 Examples of decomposable functions which are at the same time aggregation functions on L^I and belong also to $\mathscr{A}_\pi$ and $\mathscr{A}_\nu$ are projections $\mathscr{P}_1$, $\mathscr{P}_2$. Let $\mathbf{x} = [\underline{x}, \overline{x}]$, $\mathbf{y} = [\underline{y}, \overline{y}]$, then $\mathscr{P}_1(\mathbf{x}, \mathbf{y}) = [\underline{x}, \overline{x}]$, $\mathscr{P}_2(\mathbf{x}, \mathbf{y}) = [\underline{y}, \overline{y}]$. Note that they are also pseudomax A_1A_2-representable aggregation functions (with $A_1 = A_2 = P_1$ in (2.37) for $\mathscr{P}_1$ and $A_1 = A_2 = P_2$ in (2.37) for $\mathscr{P}_2$). Similarly, $\mathscr{P}_1$ and $\mathscr{P}_2$ are pseudomin A_1A_2-representable aggregation functions (they belong to the family $\mathscr{A}^*$).

Example 2.27 Let $\mathbf{x} = [\underline{x}, \overline{x}]$, $\mathbf{y} = [\underline{y}, \overline{y}]$. An example of a function which is at the same time a decomposable aggregation function on L^I and belongs also to $\mathscr{A}_\pi$ and $\mathscr{A}_\nu$ but it does not belong to the family $\mathscr{A}^*$ is $\mathscr{A}(\mathbf{x}, \mathbf{y}) = [\underline{x}\underline{y}, \overline{x}\overline{y}]$.

Theorem 2.8 ([61]) *Let $F_1, F_2 : [0, 1]^2 \to [0, 1]$ be components of a decomposable operator $\mathscr{F} : (L^I)^2 \to L^I$. If $F_1 \leqslant F_2$, F_1 is an aggregation function and $F_2(0, 0) = 0$, $F_2(1, 1) = 1$ (analogously $F_1 \leqslant F_2$, $F_1(0, 0) = 0$, $F_1(1, 1) = 1$, F_2 is an aggregation function), then $\mathscr{F}$ is a (decomposable) pos-aggregation function.*

Corollary 2.2 ([61]) *If $\mathscr{F} : (L^I)^2 \to L^I$ is a decomposable aggregation function on L^I, then $\mathscr{F}$ is a decomposable pos-aggregation function.*

The converse statement to Corollary 2.2 does not hold (certainly the same is true for Theorem 2.8). Namely, operation (2.57) in Example 2.29 shows that $\mathscr{F}$ need not be increasing and operation (2.49) in Example 2.28 shows that $\mathscr{F}$ need not be decomposable.

Example 2.28 Let $\mathbf{x} = [\underline{x}, \overline{x}]$, $\mathbf{y} = [\underline{y}, \overline{y}]$ and $A : [0, 1]^2 \to [0, 1]$ be an aggregation function, where for operations (2.49) and (2.50) let A have zero element 1, for operations (2.51) and (2.52) let A have zero element 0. The following functions are aggregation functions on L^I (non-decomposable) and they are also pos-aggregation functions (but they are not nec-aggregation functions and operations (2.53) and (2.54) do not belong to $\mathscr{A}^*$):

$$\mathscr{A}_1(\mathbf{x}, \mathbf{y}) = \begin{cases} [1, 1], & (\mathbf{x}, \mathbf{y}) = ([1, 1], [1, 1]) \\ [0, A(\underline{x}, \overline{y})], & \text{otherwise} \end{cases} \tag{2.49}$$

$$\mathscr{A}_2(\mathbf{x}, \mathbf{y}) = \begin{cases} [1, 1], & (\mathbf{x}, \mathbf{y}) = ([1, 1], [1, 1]) \\ [0, A(\overline{x}, \underline{y})], & \text{otherwise} \end{cases} \tag{2.50}$$

$$\mathscr{A}_3(\mathbf{x}, \mathbf{y}) = \begin{cases} [0, 0], & (\mathbf{x}, \mathbf{y}) = ([0, 0], [0, 0]) \\ [A(\underline{x}, \overline{y}), 1], & \text{otherwise} \end{cases} \tag{2.51}$$

$$\mathscr{A}_4(\mathbf{x}, \mathbf{y}) = \begin{cases} [0, 0], & (\mathbf{x}, \mathbf{y}) = ([0, 0], [0, 0]) \\ [A(\overline{x}, \underline{y}), 1], & \text{otherwise} \end{cases} \tag{2.52}$$

$$\mathscr{A}_5(\mathbf{x}, \mathbf{y}) = \begin{cases} [1, 1], & (\mathbf{x}, \mathbf{y}) = ([1, 1], [1, 1]) \\ \left[\frac{\underline{y}\frac{\underline{x}+\overline{x}}{2}}{2}, \frac{\overline{x}+\overline{y}}{2}\right], & \text{otherwise} \end{cases} \tag{2.53}$$

$$\mathscr{A}_6(\mathbf{x}, \mathbf{y}) = \begin{cases} [1, 1], & (\mathbf{x}, \mathbf{y}) = ([1, 1], [1, 1]) \\ \left[\frac{\underline{x}\frac{\underline{y}+\overline{y}}{2}}{2}, \frac{\overline{x}+\overline{y}}{2}\right], & \text{otherwise} \end{cases} \tag{2.54}$$

The following functions are decomposable aggregation functions on L^I, they belong to $\mathscr{A}^*$, and they are also pos-aggregation functions (but they are not nec-aggregation functions), where $A : [0, 1]^2 \to [0, 1]$ is an aggregation function:

$$\mathscr{A}(\mathbf{x}, \mathbf{y}) = \begin{cases} [1, 1], & (\mathbf{x}, \mathbf{y}) = ([1, 1], [1, 1]) \\ [0, A(\overline{x}, \overline{y})], & \text{otherwise.} \end{cases} \tag{2.55}$$

Moreover operation (2.55) may be presented as $\mathscr{A}(\mathbf{x}, \mathbf{y}) = [A_w(\underline{x}, \underline{y}), A(\overline{x}, \overline{y})]$ (A_w is the weakest aggregation function on [0, 1]) and

$$\mathscr{A}(\mathbf{x}, \mathbf{y}) = \begin{cases} [0, 0], & (\mathbf{x}, \mathbf{y}) = ([0, 0], [0, 0]) \\ [A(\underline{x}, \underline{y}), 1], & \text{otherwise,} \end{cases} \tag{2.56}$$

where operation (2.56) may be presented as $\mathscr{A}(\mathbf{x}, \mathbf{y}) = [A(\underline{x}, \underline{y}), A_s(\overline{x}, \overline{y})]$ (A_s is the strongest aggregation function on [0, 1]).

Example 2.29 Let $\mathbf{x} = [\underline{x}, \overline{x}]$, $\mathbf{y} = [\underline{y}, \overline{y}]$. The following functions on L^I are pos-aggregation functions but they are neither aggregation functions (cf. Definition 2.15) nor nec-aggregation functions (in (2.57) and (2.58) for the Lehmer mean we use the convention $\frac{0}{0} = 0$):

$$\mathscr{A}(\mathbf{x}, \mathbf{y}) = \begin{cases} [0, 0], & (\mathbf{x}, \mathbf{y}) = ([0, 0], [0, 0]) \\ \left[\frac{\underline{x}^2+\underline{y}^2}{\underline{x}+\underline{y}}, 1\right], & \text{otherwise} \end{cases} \tag{2.57}$$

$$\mathscr{A}(\mathbf{x}, \mathbf{y}) = \begin{cases} [1, 1], & (\mathbf{x}, \mathbf{y}) = ([1, 1], [1, 1]) \\ \left[0, \frac{\overline{x}^2+\overline{y}^2}{\overline{x}+\overline{y}}\right], & \text{otherwise} \end{cases} \tag{2.58}$$

$$\mathscr{A}(\mathbf{x}, \mathbf{y}) = [\underline{x} \cdot |2\underline{y} - 1|, \overline{x}] \tag{2.59}$$

$$\mathscr{A}(\mathbf{x}, \mathbf{y}) = [\underline{x} \cdot |2\underline{x} - 1|, \overline{x}] \tag{2.60}$$

$$\mathscr{A}(\mathbf{x}, \mathbf{y}) = [\underline{x} \cdot |2\overline{x} - 1|, \overline{x}] \tag{2.61}$$

$$\mathscr{A}(\mathbf{x}, \mathbf{y}) = [\underline{x} \cdot |2\overline{y} - 1|, \overline{x}] \tag{2.62}$$

$$\mathscr{A}(\mathbf{x}, \mathbf{y}) = [\underline{x} - (\max(0, \underline{x} - \underline{y}))^2, \overline{x}] \tag{2.63}$$

$$\mathscr{A}(\mathbf{x}, \mathbf{y}) = [\underline{x} - (\max(0, \underline{x} - \overline{y}))^2, \overline{x}], \tag{2.64}$$

where (2.57)–(2.60) and (2.63) are decomposable operations.

Other examples of proper functions from the family $\mathscr{A}_\pi$ may be created in a similar way as shown above, where instead of the Lehmer mean (or the weighted Lehmer mean) in Example 2.29 we may use a non-increasing, in a classical way, function. It was done for operations (2.59)–(2.64) by applying $F(x, y) = x - (\max(0, x - y))^2$ or $F(x, y) = x \cdot |2y - 1|$ in the first coordinate. Such family of functions consists of pre-aggregation functions (cf. Definition 2.13).

There exist decomposable operations $\mathscr{F} : (L^I)^2 \to L^I$ which are pos-aggregation functions but they are not aggregation functions (cf. Example 2.29). That is not the case for nec-aggregation functions which is shown by Theorem 2.9, which is the characterization of decomposable nec-aggregation functions.

Theorem 2.9 ([61]) *Let $\mathscr{F} : (L^I)^2 \to L^I$ be a decomposable operation. $\mathscr{F}$ is a nec-aggregation function if and only if $F_1 = F_2$, F_1 is an aggregation function on* [0, 1].

A decomposable operation $\mathscr{F}$ may be at the same time an aggregation function, a pos-aggregation function and a nec-aggregation function. Examples of such operations are $\mathscr{A}_s$ and $\mathscr{A}_w$, which have been already presented in Proposition 2.5. Moreover, by Corollary 2.2 and Theorem 2.9 we obtain the following result.

Corollary 2.3 ([61]) *If $\mathscr{F} : (L^I)^2 \to L^I$ is a decomposable nec-aggregation function, then $\mathscr{F}$ is a decomposable aggregation function. Moreover, if $\mathscr{F} : (L^I)^2 \to L^I$ is a decomposable nec-aggregation function, then $\mathscr{F}$ is a decomposable pos-aggregation function.*

The converse statement to Corollary 2.3 is not true which is shown by the next example.

Example 2.30 Let $\mathbf{x} = [\underline{x}, \overline{x}], \mathbf{y} = [\underline{y}, \overline{y}] \in L^I$ and a decomposable aggregation function on L^I be given by the formula $\mathscr{A}(\mathbf{x}, \mathbf{y}) = [\sqrt{\underline{x}\underline{y}}, \frac{\overline{x}+\overline{y}}{2}]$. $\mathscr{A}$ is a pos-aggregation function and it is not a nec-aggregation function.

Now we will consider non-decomposable aggregation functions on L^I and their connections with the families $\mathscr{A}_\nu$ and $\mathscr{A}_\pi$.

Proposition 2.6 ([61]) *If $A_1 = A_2$, then pseudomax A_1A_2-representable aggregation function $\mathscr{A}$ is a nec-aggregation function.*

If $A_1 < A_2$ we may not have at the same time pseudomax A_1A_2-representable aggregation function and a nec-aggregation function.

Example 2.31 Let $\mathbf{x} = [\underline{x}, \overline{x}], \mathbf{y} = [\underline{y}, \overline{y}] \in L^I$ and $\mathscr{A}(\mathbf{x}, \mathbf{y}) = [\underline{x}\underline{y}, \max(\frac{\underline{x}+\overline{y}}{2}, \frac{\overline{x}+\underline{y}}{2})]$. $\mathscr{A}$ is a pseudomax A_1A_2-representable aggregation function but $\mathscr{A} \notin \mathscr{A}_\nu$. Moreover, $\mathscr{A} \notin \mathscr{A}_\pi$.

Similarly, we may prove that

Proposition 2.7 ([61]) *If $A_1 = A_2$, then pseudomin A_1A_2-representable aggregation function $\mathscr{A}$ is a nec-aggregation function.*

If $A_1 < A_2$ we may not have at the same time pseudomin A_1A_2-representable aggregation function and a nec-aggregation function.

Example 2.32 Let $\mathbf{x} = [\underline{x}, \overline{x}], \mathbf{y} = [\underline{y}, \overline{y}] \in L^I$ and $\mathscr{A}(\mathbf{x}, \mathbf{y}) = [\min(\underline{x}\overline{y}, \overline{x}\underline{y}), \frac{\overline{x}+\overline{y}}{2}]$. $\mathscr{A}$ is a pseudomin A_1A_2-representable aggregation function but it is not a nec-aggregation function. Moreover, $\mathscr{A} \in \mathscr{A}_\pi$.

Proposition 2.8 ([61]) *If $A_1 = A_2$ and $A_3 = A_4$ in* (2.40)*, then $\mathscr{A} \in \mathscr{A}_\nu$.*

Example 2.33 If $A_1 < A_2$ and $A_3 < A_4$ in (2.40), then $\mathscr{A}$ may not be a nec-aggregation function (for example if $A_1 = A_3$ is the product and $A_2 = A_4$ is the arithmetic mean, moreover $\mathscr{A} \notin \mathscr{A}_\pi$).

Example 2.34 Let $\mathbf{x} = [\underline{x}, \overline{x}], \mathbf{y} = [\underline{y}, \overline{y}] \in L^I$. Thus operation $\mathscr{A}(\mathbf{x}, \mathbf{y}) = [\underline{x}\overline{x}, \underline{x}]$ is a non-decomposable aggregation function on L^I, which belongs to $\mathscr{A}^*$ (it fulfils (2.40) with $A_1 = A_2 = P_1$, $A_3 = T_P$, $A_4 = P_1$), but $\mathscr{A}$ is not a nec-aggregation function and $\mathscr{A}$ is not a pos-aggregation function.

Now, we will consider connections between the family of aggregation operators $\mathscr{A}_\pi$ and aggregation operators from $\mathscr{A}^*$ (cf. Example 2.15 and Definition 2.19). Even if $A_1 = A_2$, then a pseudomin A_1A_2-representable aggregation function (a pseudomax A_1A_2-representable aggregation function) may not be a pos-aggregation function. This is shown by the next two examples.

Example 2.35 Let $\mathbf{x} = [\underline{x}, \overline{x}], \mathbf{y} = [\underline{y}, \overline{y}] \in L^I$ and $\mathscr{A}(\mathbf{x}, \mathbf{y}) = [\underline{x}\underline{y}, \max(\underline{x}\overline{y}, \overline{x}\underline{y})]$. Thus $\mathscr{A}$ is a pseudomax A_1A_2-representable aggregation function but it is not a pos-aggregation function. Similar conclusion we get for $\mathscr{A}(\mathbf{x}, \mathbf{y}) = \left[\frac{\underline{x}+\underline{y}}{2}, \max(\frac{\underline{x}+\overline{y}}{2}, \frac{\overline{x}+\underline{y}}{2})\right]$.

And vice versa, a pos-aggregation function (2.54) is not a pseudomax A_1A_2-representable aggregation function.

Example 2.36 Let $\mathbf{x} = [\underline{x}, \overline{x}], \mathbf{y} = [\underline{y}, \overline{y}] \in L^I$ and $\mathscr{A}(\mathbf{x}, \mathbf{y}) = [\min(\underline{x}\overline{y}, \overline{x}\underline{y}), \overline{x}\overline{y}]$. Thus $\mathscr{A}$ is a pseudomin A_1A_2-representable aggregation function but it is not a pos-aggregation function.

And vice versa, a pos-aggregation function (2.53) is not a pseudomin A_1A_2–representable aggregation function.

Example 2.37 For $A_1 = A_2 = T_P$, we see that an aggregation operator of the form (2.39) (and certainly an aggregation operator of the form (2.40)) may not be a pos-aggregation function.

And vice versa, pos-aggregation functions (2.53) and (2.54) are not aggregation operators of the form (2.39) (and of the form (2.40)).

However, we may also consider the following operation

Example 2.38 Let $\mathbf{x} = [\underline{x}, \overline{x}], \mathbf{y} = [\underline{y}, \overline{y}] \in L^I$. A non-decomposable interval-valued aggregation function $\mathcal{A}(\mathbf{x}, \mathbf{y}) = [\underline{x}\overline{x}, \overline{x}]$ fulfils the condition $\mathcal{A} \in \mathcal{A}^*$ (it is of the form (2.40)), $\mathcal{A} \in \mathcal{A}_\pi$ but $\mathcal{A} \notin \mathcal{A}_\nu$.

Now we will give examples of aggregation operators belonging to $\mathcal{A}_\nu$ which may or may not belong to other considered here families of functions. Especially, these functions are not decomposable ones and they are not of the form (2.38)–(2.39).

Example 2.39 Let $\mathbf{x} = [\underline{x}, \overline{x}], \mathbf{y} = [\underline{y}, \overline{y}] \in L^I$. The following interval-valued aggregation function $\mathcal{A}(\mathbf{x}, \mathbf{y}) = [\underline{x}\overline{x}, \underline{x}\overline{x}]$ is not decomposable and $\mathcal{A} \in \mathcal{A}^*$ (it is enough to put $A_1 = A_2 = P_1$ and $A_3 = A_4 = T_P$ in (2.40)), $\mathcal{A} \in \mathcal{A}_\nu$ but $\mathcal{A} \notin \mathcal{A}_\pi$.

Example 2.40 Let $\mathbf{x} = [\underline{x}, \overline{x}], \mathbf{y} = [\underline{y}, \overline{y}] \in L^I$. The following interval-valued aggregation function $\mathcal{A}(\mathbf{x}, \mathbf{y}) = \left[\frac{\underline{y}+\frac{\underline{x}+\overline{x}}{2}}{2}, \frac{\overline{x}+\overline{y}}{2}\right]$ is not decomposable, is not of the form (2.38)–(2.39), i.e. $\mathcal{A} \notin \mathcal{A}^*$ it belongs to $\mathcal{A}_\nu$, it does not belong to $\mathcal{A}_\pi$.

Example 2.41 Let $\mathbf{x} = [\underline{x}, \overline{x}], \mathbf{y} = [\underline{y}, \overline{y}] \in L^I$. $\mathcal{A}(\mathbf{x}, \mathbf{y}) = \left[\frac{\underline{x}+\overline{x}}{2}, \frac{\underline{x}^2+\overline{x}^2}{\underline{x}+\overline{x}}\right]$, is not an aggregation function on L^I and $\mathcal{A} \in \mathcal{A}_\nu$ (it is a proper nec-aggregation function) and $\mathcal{A} \notin \mathcal{A}_\pi$, $\mathcal{A} \notin \mathcal{D}$.

Below we present an example of operator which is neither aggregation function on L^I nor pos-aggregation function nor nec-aggregation function.

Example 2.42 The following function $\mathcal{A}(\mathbf{x}, \mathbf{y}) = [\overline{x} \cdot |2\overline{x} - 1|, \overline{x}]$ is not an aggregation function on L^I, $\mathcal{A} \notin \mathcal{A}_\nu$, $\mathcal{A} \notin \mathcal{A}_\pi$, $\mathcal{A} \notin \mathcal{D}$. $\mathcal{A}$ fulfils boundary conditions but it does not fulfil any of the monotonicity conditions (2.32), (2.41), (2.42).

In the next subsection dependencies between pos- and nec-aggregation functions and and interval-valued aggregation functions with respect to the admissible linear orders are provided.

2.3.2 *Aggregation Functions with Respect to Linear Orders Versus Other Classes of Aggregation Functions in Interval-Valued Fuzzy Settings*

We consider here interval-valued aggregation functions with respect to admissible linear orders presented in Sect. 2.2.3.

If some operator is an aggregation function with respect to the partial order $\preceq$ it may not be an aggregation function with respect to some admissible linear order $\leq_{L^I}$. That is the case with simple examples of aggregation functions which are lattice operations $\wedge$ and $\vee$ (cf. [69]). Let us consider $\mathscr{A}(\mathbf{x}, \mathbf{y}) = [\min(\underline{x}, \underline{y}), \min(\overline{x}, \overline{y})]$, the admissible linear order $\leq_{Lex2}$ and $\mathbf{x} = [0.3, 0.4]$, $\mathbf{y} = [0.2, 0.6]$, $\mathbf{u} = [0.1, 0.7]$. We see that $\mathbf{x} \leq_{Lex2} \mathbf{y} \leq_{Lex2} \mathbf{u}$ however, $\mathscr{A}(\mathbf{x}, \mathbf{u}) = [0.1, 0.4] \leq_{Lex2} [0.2, 0.4] = \mathscr{A}(\mathbf{x}, \mathbf{y})$. Similarly, let us consider $\mathscr{A}(\mathbf{x}, \mathbf{y}) = [\max(\underline{x}, \underline{y}), \max(\overline{x}, \overline{y})]$, the admissible linear order $\leq_{Lex1}$ and $\mathbf{x} = [0.3, 0.9]$, $\mathbf{y} = [0.4, 0.8]$, $\mathbf{u} = [0.6, 0.7]$. We see that $\mathbf{x} \leq_{Lex1} \mathbf{y} \leq_{Lex1} \mathbf{u}$ however, $\mathscr{A}(\mathbf{y}, \mathbf{u}) = [0.6, 0.8] \leq_{Lex1} [0.6, 0.9] = \mathscr{A}(\mathbf{x}, \mathbf{u})$.

If it comes to the mutual relations between aggregation functions with respect to admissible linear orders and pos- and nec-aggregation functions, there are not clear dependencies between them. Examples 2.43 and 2.44 show that the family of interval-valued aggregation functions with respect to a fixed admissible linear order may intersect the family of pos-aggregation functions but none of this families is included in the other.

Example 2.43 The operator (2.56) is an interval-valued aggregation function with respect to the linear order $\leq_{Lex1}$ (cf. [69]). This operator is also a decomposable interval-valued aggregation function which is a pos-aggregation function but it is not a nec-aggregation function (cf. Example 2.28). Similarly, the following

$$\mathscr{A}(\mathbf{x}, \mathbf{y}) = \begin{cases} [1, 1], & (\mathbf{x}, \mathbf{y}) = ([1, 1], [1, 1]) \\ [0, A(\underline{x}, \underline{y})], & \text{otherwise} \end{cases}$$

is an interval-valued aggregation function with respect to the linear order $\leq_{Lex1}$ (cf. [69]). This is a non-decomposable interval-valued aggregation function which is neither a pos-aggregation function nor a nec-aggregation function. Moreover, a pos-aggregation function (2.57) is not an interval-valued aggregation function with respect to the linear order $\leq_{Lex1}$. Let us consider $\mathbf{x} = [0.04, 1]$, $\mathbf{y} = [0.05, 1]$ and $\mathbf{u} = [0.5, 1]$. We see that $\mathbf{x} \leq_{Lex1} \mathbf{y}$ but for the aggregation operator $\mathscr{A}$ defined by (2.57) we have $[0.459, 1] = \mathscr{A}(\mathbf{y}, \mathbf{u}) \leq_{Lex1} \mathscr{A}(\mathbf{x}, \mathbf{u}) = [0.465, 1]$.

Example 2.44 The following is a non-decomposable interval-valued aggregation function but it is neither a pos-aggregation function nor a nec-aggregation function, where

$$\mathscr{A}(\mathbf{x}, \mathbf{y}) = \begin{cases} [0, 0], & (\mathbf{x}, \mathbf{y}) = ([0, 0], [0, 0]) \\ [A(\overline{x}, \overline{y}), 1], & \text{otherwise.} \end{cases}$$

This is an interval-valued aggregation function with respect to the linear order $\leq_{Lex2}$ (cf. [69]). Moreover, the operator (2.55) is an interval-valued aggregation function with respect to the linear order $\leq_{Lex2}$ (cf. [69]). As it was stated in Example 2.28 this is also a decomposable aggregation function which is simultaneously a pos-aggregation function but it is not a nec-aggregation function. Moreover, a pos-aggregation function (2.58) is not an interval-valued aggregation function with respect to the linear order $\leq_{Lex2}$. Let us consider $\mathbf{x} = [0, 0.04]$, $\mathbf{y} = [0, 0.05]$ and $\mathbf{u} = [0, 0.5]$. We see that $\mathbf{x} \leq_{Lex2} \mathbf{y}$ but for the aggregation operator $\mathscr{A}$ defined by (2.58) we have $[0, 0.459] = \mathscr{A}(\mathbf{y}, \mathbf{u}) \leq_{Lex2} \mathscr{A}(\mathbf{x}, \mathbf{u}) = [0, 0.465]$.

Similar results may be obtained for the family of nec-aggregation functions, i.e. the family of interval-valued aggregation functions with respect to a fixed admissible linear order may intersect the family of nec-aggregation functions but none of this families is included in the other. The following operator $\mathscr{A}([\underline{x}, \overline{x}], [\underline{y}, \overline{y}]) = [(\underline{x}\underline{y})^p, (\overline{x}\overline{y})^p]$ is a nec-aggregation function (cf. Theorem 2.9). It is also an interval-valued aggregation function with respect to $\leq_{Lex2}$ (cf. [62]). As a result, we see that an interval-valued aggregation function with respect to the same admissible linear order may be or not a nec-aggregation function (cf. Example 2.44). Moreover, a nec-aggregation function (cf. Proposition 2.6) $\mathscr{A}(\mathbf{x}, \mathbf{y}) = [\underline{x}\underline{y}, \max(\underline{x}\overline{y}, \overline{x}\underline{y})]$ is not an interval-valued aggregation function with respect to the linear order $\leq_{Lex2}$. Namely, let us consider $\mathbf{x} = \mathbf{u} = [0.5, 0.5]$, $\mathbf{y} = \mathbf{w} = [0, 0.6]$. We see that $\mathbf{x} \leq_{Lex2} \mathbf{y}$, $\mathbf{u} \leq_{Lex2} \mathbf{w}$ but $[0, 0] = \mathscr{A}(\mathbf{y}, \mathbf{w}) \leq_{Lex2} \mathscr{A}(\mathbf{x}, \mathbf{u}) = [0.25, 0.25]$.

There are some open interesting problems concerning dependencies between interval-valued aggregation functions with respect to admissible linear orders and other aggregation operators. We may restrict ourselves to the construction method of admissible orders presented in Proposition 1.4. First of all, we may ask if an interval-valued aggregation function with respect to an admissible linear order is also an interval-valued aggregation function with respect to the partial order $\preceq$ (to simplify the problem we may restrict this question to the case of decomposable operators, since in the case of decomposability there is a clear dependence between interval-valued aggregation functions and pos- and nec-aggregation functions, cf. Corollary 2.3). Other interesting questions are as follows. Does it exist an aggregation function with respect to the partial order (or pos-aggregation function, nec-aggregation function) such that it is not linear with respect to any of the admissible linear orders? Moreover, for a given interval-valued aggregation function (or pos-aggregation function, nec-aggregation function), is it possible to determine the family of all admissible linear orders with respect to which this aggregation operator would be linear?

2.4 Construction Methods of Aggregation Operators in Interval-Valued Fuzzy Settings

There exist diverse construction methods of interval-valued aggregation operators. Some ways of obtaining aggregation functions have been already given, for example Theorems 2.7, 2.9, Definition 2.19, formulas (2.49)–(2.52) in Example 2.28. Here, we consider other classical methods of generation of new operators from the given ones. These are aggregation methods, construction of a dual operator, composition methods. For the simplicity of notations there will be presented construction methods involving two operators but all methods may be extended to the case of a finite number of operators.

Proposition 2.9 *Let $\mathcal{B}, \mathcal{C} : (L^I)^2 \to L^I$ be pos-aggregation functions (respectively nec-aggregation functions or interval-valued aggregation functions). If $\mathcal{A} : (L^I)^2 \to L^I$ is a representable interval-valued aggregation function, then $\mathcal{A}(\mathcal{B}, \mathcal{C})$ is a pos-aggregation function (respectively nec-aggregation function or interval-valued aggregation function).*

Proof We consider the case of pos-aggregation functions. The remaining dependencies may be proven analogously. Let $\mathbf{x} = [\underline{x}, \overline{x}]$, $\mathbf{y} = [\underline{y}, \overline{y}]$, $\mathbf{u} = [\underline{u}, \overline{u}]$, $\mathbf{w} = [\underline{w}, \overline{w}]$ and $\mathbf{x} \preceq_\pi \mathbf{y}$, $\mathbf{u} \preceq_\pi \mathbf{w}$. We will show that if $\mathcal{A}$ has component aggregation functions A_1, A_2 on $[0, 1]$ such that $A_1 \leqslant A_2$, then $\mathcal{A}(\mathcal{B}, \mathcal{C})(\mathbf{x}, \mathbf{u}) \preceq_\pi \mathcal{A}(\mathcal{B}, \mathcal{C})(\mathbf{y}, \mathbf{w})$. Since $\mathcal{B}$ and $\mathcal{C}$ are pos-aggregation functions, then with the given assumptions we get $\underline{\mathcal{B}}(\mathbf{x}, \mathbf{u}) \leqslant \overline{\mathcal{B}}(\mathbf{y}, \mathbf{w})$ and $\underline{\mathcal{C}}(\mathbf{x}, \mathbf{u}) \leqslant \overline{\mathcal{C}}(\mathbf{y}, \mathbf{w})$. By monotonicity of A_1, A_2 and since $A_1 \leqslant A_2$, we have

$$A_1(\underline{\mathcal{B}}(\mathbf{x}, \mathbf{u}), \underline{\mathcal{C}}(\mathbf{x}, \mathbf{u})) \leqslant A_2(\underline{\mathcal{B}}(\mathbf{x}, \mathbf{u}), \underline{\mathcal{C}}(\mathbf{x}, \mathbf{u})) \leqslant A_2(\overline{\mathcal{B}}(\mathbf{y}, \mathbf{w}), \overline{\mathcal{C}}(\mathbf{y}, \mathbf{w})).$$

The boundary conditions are fulfilled by definition of $\mathcal{A}(\mathcal{B}, \mathcal{C})$. This finishes the proof.

Let us note that in Proposition 2.9 there is no need to consider decomposable pos-aggregation functions or decomposable nec-aggregation functions $\mathcal{A}$. In the proof we used assumptions of boundary conditions and monotonicity of A_1 and A_2. By Theorem 2.7 it means that such operator is necessarily an interval-valued aggregation function. Moreover, Proposition 2.9 is fulfilled for arbitrary aggregation operators $\mathcal{B}$ and $\mathcal{C}$, not necessarily decomposable ones.

If $A_1 = A_2$, by Proposition 2.9 and Theorem 2.9 we get the following result.

Corollary 2.4 *Let $\mathcal{B}, \mathcal{C} : (L^I)^2 \to L^I$ be pos-aggregation functions (respectively nec-aggregation functions or interval-valued aggregation functions). If $\mathcal{A} : (L^I)^2 \to L^I$ is a decomposable nec-aggregation function, then $\mathcal{A}(\mathcal{B}, \mathcal{C})$ is a pos-aggregation function (respectively nec-aggregation function or interval-valued aggregation function).*

Now, we will consider the notion of a $\mathcal{N}$-dual operator to a given one with respect to some interval-valued fuzzy negation $\mathcal{N}$.

Definition 2.24 Let $\mathcal{F} : L^I \to L^I$ be a decomposable operator with the components $F_1, F_2 : [0, 1]^2 \to [0, 1]$, $\mathcal{N}$ be a strong representable interval-valued fuzzy negation with the associate strong fuzzy negation N. $\mathcal{F}^{\mathcal{N}}$ is called a $\mathcal{N}$-dual operator to the operator $\mathcal{F}$, where

$$\mathcal{F}^{\mathcal{N}}(\mathbf{x}, \mathbf{y}) = \mathcal{N}(\mathcal{F}(\mathcal{N}(\mathbf{x}), \mathcal{N}(\mathbf{y}))), \ \mathbf{x} = [\underline{x}, \overline{x}], \mathbf{y} = [\underline{y}, \overline{y}] \in L^I.$$

By decomposability of $\mathcal{F}$ operator $\mathcal{F}^{\mathcal{N}}$ has the following representation

$$\mathcal{F}^{\mathcal{N}}(\mathbf{x}, \mathbf{y}) = [N(F_2(N(\underline{x}), N(\underline{y}))), N(F_1(N(\overline{x}), N(\overline{y})))], \mathbf{x}, \mathbf{y} \in L^I. \quad (2.65)$$

Operator (2.65) is well defined. Namely, if $\mathcal{F}$ is a well defined decomposable operator, then $F_1(\underline{x}, \underline{y}) \leqslant F_2(\overline{x}, \overline{y})$ for any intervals $[\underline{x}, \overline{x}], [\underline{y}, \overline{y}] \in L^I$ (where certainly $\underline{x} \leqslant \overline{x}$, $\underline{y} \leqslant \overline{y}$). As a result for a strong fuzzy negation N, associated with a strong representable interval-valued fuzzy negation $\mathcal{N}$, we get the adequate intervals $[N(\overline{x}), N(\underline{x})], [N(\overline{y}), N(\underline{y})] \in L^I$. Since $\mathcal{F}$ is a well defined decomposable operator, $F_1(N(\overline{x}), N(\overline{y})) \leqslant F_2(N(\underline{x}), N(\underline{y}))$ and we get $N(F_2(N(\underline{x}), N(\underline{y})) \leqslant N(F_1(N(\overline{x}), N(\overline{y})))$.

$\mathcal{N}$-dual operator to a pseudomin representable aggregation function or a pseudomax representable aggregation function may not be a properly defined operator (cf. Example 2.45). This is why we assume that $\mathcal{N}$-dual operator with respect to Definition 2.24 is defined for decomposable operators.

Example 2.45 Let $\mathcal{A}(\mathbf{x}, \mathbf{y}) = [\underline{x}\underline{y}, \max(\frac{\underline{x}+\overline{y}}{2}, \frac{\overline{x}+\underline{y}}{2})]$, where $\mathbf{x} = [\underline{x}, \overline{x}]$, $\mathbf{y} = [\underline{y}, \overline{y}] \in L^I$. The corresponding $\mathcal{N}$-dual operator to $\mathcal{A}$, with respect to an associate strong fuzzy negation $N = 1 - x$ (the classical fuzzy negation), is of the form $\mathcal{A}^{\mathcal{N}}(\mathbf{x}, \mathbf{y}) = [\min(\underline{x} + \overline{y}, \overline{x} + \underline{y}), \overline{x} + \overline{y} - \overline{x}\overline{y}]$. If $\underline{x} = \overline{x} = 0.3$ and $\underline{y} = \overline{y} = 0.9$, then $\mathcal{A}^{\mathcal{N}}(\mathbf{x}, \mathbf{y}) = [1.2, 0.93]$. We see that for the considered pair $\mathbf{x}, \mathbf{y}$ it holds $\underline{\mathcal{A}^{\mathcal{N}}}(\mathbf{x}, \mathbf{y}) > \overline{\mathcal{A}^{\mathcal{N}}}(\mathbf{x}, \mathbf{y})$. As a result, for the given $\mathcal{A}$ (which is a pseudomax representable aggregation function) there is no sense to apply Definition 2.24.

Proposition 2.10 *If $\mathcal{F} : L^I \to L^I$ is a decomposable interval-valued aggregation function (decomposable pos-aggregation function or decomposable nec-aggregation function, respectively), then $\mathcal{F}^{\mathcal{N}}$ is also a decomposable interval-valued aggregation function (decomposable pos-aggregation function or decomposable nec-aggregation function, respectively).*

Proof We justify the property for decomposable pos-aggregation functions. Other operators may be be considered similarly. Let $\mathcal{N}$ be a strong interval-valued fuzzy negation with its associate strong fuzzy negation N, $\mathbf{x} = [\underline{x}, \overline{x}]$, $\mathbf{y} = [\underline{y}, \overline{y}]$, $\mathbf{u} = [\underline{u}, \overline{u}]$, $\mathbf{w} = [\underline{w}, \overline{w}]$ and $\mathbf{x} \preceq_\pi \mathbf{y}$, $\mathbf{u} \preceq_\pi \mathbf{w}$. We will show that if $\mathcal{F}$ has component functions F_1, F_2, then $\mathcal{F}^{\mathcal{N}}(\mathbf{x}, \mathbf{u}) \preceq_\pi \mathcal{F}^{\mathcal{N}}(\mathbf{y}, \mathbf{w})$. By the given assumptions $\underline{x} \leqslant \overline{y}$ and $\underline{u} \leqslant \overline{w}$. Thus $N(\overline{y}) \leqslant N(\underline{x})$ and $N(\overline{w}) \leqslant N(\underline{u})$ and we get adequate intervals $[N(\overline{x}), N(\underline{x})]$, $[N(\overline{y}), N(\underline{y})]$, $[N(\overline{u}), N(\underline{u})]$, $[N(\overline{w}), N(\underline{w})]$. As a result $[N(\overline{y}), N(\underline{y})] \preceq_\pi [N(\overline{x}), N(\underline{x})]$ and we have $[N(\overline{w}), N(\underline{w})] \preceq_\pi [N(\overline{u}), N(\underline{u})]$. Since $\mathcal{F}$ is a pos-aggregation function, $F_1(N(\overline{y}), N(\overline{w})) \leqslant F_2(N(\underline{x}), N(\underline{u}))$. By monotonicity of N we get $N(F_2(N(\underline{x}), N(\underline{u}))) \leqslant N(F_1(N(\overline{y}), N(\overline{w})))$. This means that $\mathcal{F}^{\mathcal{N}}(\mathbf{x}, \mathbf{u}) \preceq_\pi \mathcal{F}^{\mathcal{N}}(\mathbf{y}, \mathbf{w})$, where $\mathcal{F}^{\mathcal{N}}(\mathbf{x}, \mathbf{u}) = [N(F_2(N(\underline{x}), N(\underline{u}))), N(F_1(N(\overline{x}), N(\overline{u})))]$ and $\mathcal{F}^{\mathcal{N}}(\mathbf{y}, \mathbf{w}) = [N(F_2(N(\underline{y}), N(\underline{w}))), N(F_1(N(\overline{y}), N(\overline{w})))]$. The boundary conditions are fulfilled by definition of $\mathcal{F}^{\mathcal{N}}$. This finishes the proof.

Proposition 2.11 *Let $\mathcal{F} : L^I \to L^I$ be a decomposable operator, $\mathcal{N}$ be a representable interval-valued fuzzy negation with the associate strong fuzzy negation N. $\mathcal{F}$ is self $\mathcal{N}$-dual if and only if its component functions $F_1, F_2 : [0, 1]^2 \to [0, 1]$ are N-dual to each other.*

Proof We have to show that $\mathscr{F}^{\mathscr{N}} = \mathscr{F}$ if and only if $F_1^N = F_2$ (this means that also $F_2^N = F_1$). Let $\mathbf{x}, \mathbf{y} \in L^I$. By (2.65) and Definitions 2.12, 2.24 we see that

$$[N(F_2(N(\underline{x}), N(\underline{y}))), N(F_1(N(\overline{x}), N(\overline{y})))] = [F_1(\underline{x}, \underline{y}), F_2(\overline{x}, \overline{y})]$$

is equivalent to $F_1^N = F_2$.

Corollary 2.5 *Let $\mathscr{N}$ be a representable interval-valued fuzzy negation with the associate strong fuzzy negation N and $\mathscr{A} : L^I \to L^I$ be a decomposable interval-valued aggregation function (a decomposable pos-aggregation function or a decomposable nec-aggregation function). $\mathscr{A}$ is self $\mathscr{N}$-dual if and only if its component functions $A_1, A_2 : [0,1]^2 \to [0,1]$ are N-dual to each other.*

Examples of self $\mathscr{N}$-dual decomposable operators $\mathscr{F}$ provide N-dual to each other components of $\mathscr{F}$. These are for example conjunctions and disjunctions (including t-norms and t-conorms). Let us consider, as an example, the following operator $\mathscr{F}(\mathbf{x}, \mathbf{y}) = [\underline{x} \cdot \underline{y}, \overline{x} + \overline{y} - \overline{x} \cdot \overline{y}]$, where $F_1 = T_P$ and $F_2 = S_P$, $N(x) = 1 - x$. Another example of a self $\mathscr{N}$-dual operator is the representable arithmetic mean (cf. Example 2.12), where the associate fuzzy negation is the classical fuzzy negation, i.e. $N(x) = 1 - x$ and $F_1^N = F_1$ is the arithmetic mean. Other examples of such operators may be constructed with the use of Example 2.8.

Another construction method of aggregation operators for interval-valued fuzzy settings is a composition defined for bounds of intervals.

Proposition 2.12 *Let $\mathscr{A}$ be a pos-aggregation function (a nec-aggregation function or an interval-valued aggregation function), $\varphi_1, \varphi_2 : [0,1] \to [0,1]$ be increasing bijections such that $\varphi_1 \leqslant \varphi_2$. Thus operator $\mathscr{B}(\mathbf{x}, \mathbf{y}) = [\varphi_1(\underline{\mathscr{A}}(\mathbf{x}, \mathbf{y})), \varphi_2(\overline{\mathscr{A}}(\mathbf{x}, \mathbf{y}))]$ for $\mathbf{x}, \mathbf{y} \in L^I$ is also a pos-aggregation function (a nec-aggregation or an interval-valued aggregation function, respectively).*

Proof Boundary conditions for operator $\mathscr{B}$ are fulfilled by assumptions on φ_1, φ_2, since $\varphi_1(0) = \varphi_2(0) = 0$ and $\varphi_1(1) = \varphi_2(1) = 1$. Conditions of monotonicity of $\mathscr{B}$ are fulfilled by monotonicity of φ_1, φ_2.

Let us note that Proposition 2.12 is fulfilled for arbitrary operators, not necessarily decomposable ones. In Proposition 2.12 it is not possible to consider decreasing bijections φ_1, φ_2, since for such functions boundary conditions for $\mathscr{B}$ will be not fulfilled.

2.5 Properties of Aggregation Functions in Interval-Valued Fuzzy Settings

In this section there are presented properties of operators (cf. [11, 70]) for interval-valued fuzzy settings with special attention paid to possible and necessary aggregation functions. The most commonly used properties are examined. For simplicity of notions binary versions of the considered concepts are presented. Here we focus on the problem if an interval-valued operator preserves properties of its components, i.e. inheritance of a given property by an interval-valued operator from its components is considered.

Definition 2.25 (*cf.* [11, 70]) Let $\mathbf{x} = [\underline{x}, \overline{x}], \mathbf{y} = [\underline{y}, \overline{y}], \mathbf{z} = [\underline{z}, \overline{z}], \mathbf{t} = [\underline{t}, \overline{t}] \in L^I$. Thus an operator $\mathscr{F} : (L^I)^2 \to L^I$:

- is symmetric, if

$$\mathscr{F}([\underline{x}, \overline{x}], [\underline{y}, \overline{y}]) = \mathscr{F}([\underline{y}, \overline{y}], [\underline{x}, \overline{x}]),$$

- is associative, if

$$\mathscr{F}(\mathscr{F}([\underline{x}, \overline{x}], [\underline{y}, \overline{y}]), [\underline{z}, \overline{z}]) = \mathscr{F}([\underline{x}, \overline{x}], \mathscr{F}([\underline{y}, \overline{y}], [\underline{z}, \overline{z}])),$$

- is bisymmetric, if

$$\mathscr{F}(\mathscr{F}([\underline{x}, \overline{x}], [\underline{y}, \overline{y}]), \mathscr{F}([\underline{z}, \overline{z}], [\underline{t}, \overline{t}])) = \mathscr{F}(\mathscr{F}([\underline{x}, \overline{x}], [\underline{z}, \overline{z}]), \mathscr{F}([\underline{y}, \overline{y}], [\underline{t}, \overline{t}])),$$

- is idempotent, if

$$\mathscr{F}([\underline{x}, \overline{x}], [\underline{x}, \overline{x}]) = [\underline{x}, \overline{x}],$$

- has a neutral element $[\underline{e}, \overline{e}] \in L^I$, if

$$\mathscr{F}([\underline{x}, \overline{x}], [\underline{e}, \overline{e}]) = \mathscr{F}([\underline{e}, \overline{e}], [\underline{x}, \overline{x}]) = [\underline{x}, \overline{x}],$$

- has a zero element $[\underline{a}, \overline{a}] \in L^I$, if

$$\mathscr{F}([\underline{x}, \overline{x}], [\underline{a}, \overline{a}]) = \mathscr{F}([\underline{a}, \overline{a}], [\underline{x}, \overline{x}]) = [\underline{a}, \overline{a}].$$

Properties of aggregation functions have practical interpretations. For example, aggregation methods are applied in decision making processes (cf. Chap. 3), where we have the following interpretations of the properties listed in Definition 2.25. Idempotency (in decision making called also *unanimity*) means that if all criteria are satisfied in the same degree $\mathbf{x}$, then also the global score should be $\mathbf{x}$ (cf. [25], p. 10). Symmetry reflects the same importance of a single criteria in multicriteria decision making, i.e. the knowledge of order of inputs score is irrelevant (this property is also called *anonymity*, cf. [25], p. 15). Bisymmetry generalizes simultaneous associativity and symmetry, since associative and symmetric operator is necessarily bisymmetric (cf. [25], p. 17). If an operator has a neutral element, then assigning a score equal to the neutral element (if it exists) to some criterion means that only the other criteria fulfillments are decisive for the global evaluation (cf. [25], p. 17). Whereas if an operator has a zero element $\mathbf{z}$ it means that $\mathbf{z}$ is the *veto* element.

Now, we proceed to the results connected with preservation by an interval-valued operator the properties of its components. Mainly, the results for decomposable operations will be presented.

Theorem 2.10 (cf. [56, 71]) *Let $\mathcal{F}$ be a decomposable operator, $\mathcal{F}([\underline{x}, \overline{x}], [\underline{y}, \overline{y}]) = [F_1(\underline{x}, \underline{y}), F_2(\overline{x}, \overline{y})]$ for every $[\underline{x}, \overline{x}], [\underline{y}, \overline{y}] \in L^I$ and $F_1, F_2 : [0,1]^2 \to [0,1]$.*

- *$\mathcal{F}$ is symmetric if and only if F_1, F_2 are symmetric.*
- *$\mathcal{F}$ is associative if and only if F_1, F_2 are associative.*
- *$\mathcal{F}$ is bisymmetric if and only if F_1, F_2 are bisymmetric.*
- *$\mathcal{F}$ is idempotent if and only if F_1, F_2 are idempotent.*
- *$\mathcal{F}$ has a neutral element $[\underline{e}, \overline{e}]$ if and only if $\underline{e}$ and $\overline{e}$ are neutral elements for F_1 and F_2, respectively.*
- *$\mathcal{F}$ has a zero element $[\underline{z}, \overline{z}]$ if and only if $\underline{z}$ and $\overline{z}$ are zero elements for F_1 and F_2, respectively.*

Theorem 2.10 refers to arbitrary decomposable operations as a result it refers also to decomposable pos- and nec-aggregation functions. For decomposable nec-aggregation functions we have the following results.

Corollary 2.6 *Let $\mathcal{F}$ be a decomposable nec-aggregation function, $\mathcal{F}([\underline{x}, \overline{x}], [\underline{y}, \overline{y}]) = [F(\underline{x}, \underline{y}), F(\overline{x}, \overline{y})]$ for every $[\underline{x}, \overline{x}], [\underline{y}, \overline{y}] \in L^I$ and some aggregation function $F : [0,1]^2 \to [0,1]$.*

- *$\mathcal{F}$ is symmetric if and only if F is symmetric.*
- *$\mathcal{F}$ is associative if and only if F is associative.*
- *$\mathcal{F}$ is bisymmetric if and only if F is bisymmetric.*
- *$\mathcal{F}$ is idempotent if and only if F is idempotent.*
- *$\mathcal{F}$ has a neutral element $[e, e]$ if and only if e is a neutral element of F.*
- *$\mathcal{F}$ has a zero element $[z, z]$ if and only if z is a zero element of F.*

Proof Justification follows directly from Theorem 2.10 and from Theorem 2.9.

Other dependencies for decomposable operators, such as distributivity between operators, homogeneity, transitivity properties, may be characterized analogously, i.e. by adequate components of lower and upper bound of interval [56]. However, there may be some exceptions. The property of cancellation [11] defined for interval-valued operation may not be characterized in a similar way [56].

Since non-decomposable pos- and nec-aggregation functions form diverse subclasses (cf. Sect. 2.3.1), we consider only some cases of such aggregation functions. Non-decomposable operators may not preserve the given properties of component functions. For example, pseudomax representable operations do not preserve in general idempotency of their component functions.

Example 2.46 Pseudomax A_1A_2-representable aggregation function (which is also a nec-aggregation function, cf. Proposition 2.6)

$$\mathcal{A}(\mathbf{x}, \mathbf{y}) = [\min(\underline{x}, \underline{y}), \max(\min(\underline{x}, \overline{y}), \min(\overline{x}, \underline{y}))],$$

where $A_1 = A_2 = \min$, is not idempotent while minimum is an idempotent function (cf. [56]). Namely

$$\mathscr{A}(\mathbf{x}, \mathbf{x}) = [\min(\underline{x}, \underline{x}), \max(\min(\underline{x}, \overline{x}), \min(\overline{x}, \underline{x}))] = [\underline{x}, \underline{x}],$$

then for $[\underline{x}, \overline{x}]$ such that $\underline{x} < \overline{x}$ we get $\mathscr{A}(\mathbf{x}, \mathbf{x}) \neq \mathbf{x}$. As a result $\mathscr{A}$ is not idempotent.

Non-decomposable pos-aggregation functions may preserve idempotency. For example, pos-aggregation function (2.64) is idempotent just as its component functions.

By definition of symmetry (cf. Definition 2.25) we get the following result.

Proposition 2.13 *If components A_1, A_2 of a pseudomax A_1A_2-representable aggregation function (or a pseudomin A_1A_2-representable aggregation function) are symmetric, then a pseudomax A_1A_2-representable aggregation function (or a pseudomin A_1A_2-representable aggregation function) is symmetric.*

Proof Let $[\underline{x}, \overline{x}], [\underline{y}, \overline{y}] \in L^I$. We will justify the property for a pseudomax A_1A_2-representable aggregation function. By symmetry of operation max and underlying component aggregation functions A_1 and A_2 we get

$$\mathscr{A}([\underline{x}, \overline{x}], [\underline{y}, \overline{y}]) = [A_1(\underline{x}, \underline{y}), \max(A_2(\underline{x}, \overline{y}), A_2(\overline{x}, \underline{y}))] =$$

$$[A_1(\underline{x}, \underline{y}), \max(A_2(\overline{y}, \underline{x}), A_2(\underline{y}, \overline{x}))] =$$

$$[A_1(\underline{y}, \underline{x}), \max(A_2(\underline{y}, \overline{x}), A_2(\overline{y}, \underline{x}))]] = \mathscr{A}([\underline{y}, \overline{y}], [\underline{x}, \overline{x}]).$$

Proposition 2.14 *If components A_1, A_2 of a pseudomax A_1A_2-representable aggregation function (or a pseudomin A_1A_2-representable aggregation function) have a neutral element $e \in [0, 1]$, then a pseudomax A_1A_2-representable aggregation function (or a pseudomin A_1A_2-representable aggregation function) have a neutral element $[e, e] \in L^I$.*

Proof Let $[\underline{x}, \overline{x}], [\underline{y}, \overline{y}] \in L^I$. We will prove the property for a pseudomax A_1A_2-representable aggregation function. If underlying component aggregation functions A_1 and A_2 have a neutral element e, then

$$\mathscr{A}([\underline{x}, \overline{x}], [e, e]) = [A_1(\underline{x}, e), \max(A_2(\underline{x}, e), A_2(\overline{x}, e))] = [\underline{x}, \max(\underline{x}, \overline{x})] = [\underline{x}, \overline{x}].$$

Similarly, we get $\mathscr{A}([e, e], [\underline{x}, \overline{x}]) = [\underline{x}, \overline{x}]$.

Analogously we may prove the existence if a zero element for a pseudomax or a pseudomin representable aggregation function.

Proposition 2.15 *If components A_1, A_2 of a pseudomax A_1A_2-representable aggregation function (or a pseudomin A_1A_2-representable aggregation function) have a zero element $z \in [0, 1]$, then a pseudomax A_1A_2-representable aggregation function (or a pseudomin A_1A_2-representable aggregation function) have a zero element $[z, z] \in L^I$.*

As a special case of Propositions 2.13, 2.14 and 2.15, a subfamily of nec-aggregation functions, being at the same time pseudomax or pseudomin representable aggregation functions (cf. Propositions 2.6 and 2.7), preserve symmetry, neutral element or a zero element of component functions.

Associativity and bisymmetry for pseudomax or pseudomin representable aggregation functions may be obtained with additional assumptions on component functions A_1 and A_2 (associativity or bisymmetry, respectively is not sufficient). An example of a pseudomax A_1A_2-representable aggregation function which is associative and bisymmetric is $\mathscr{A}(x, y) = [\underline{x}\underline{y}, \max(\underline{x}\overline{y}, \overline{x}\underline{y})]$. This is also a necessary aggregation function (cf. Proposition 2.6).

In the next section we will consider the concept of a width of interval, its preservation by aggregation operators or estimation of the width of output interval after aggregation.

2.6 Preservation of the Width of Intervals by Aggregation Operators

The width of a given membership interval can be seen as a measure of the uncertainty or ignorance of the expert to provide the exact real-value membership degree of the element. This is why it may be important problem for some of the applications to check dependence between the width of input intervals and the width of the output interval (after aggregation). In this section we discuss the problem whether a given aggregation operator preserves the width of input intervals or we will provide/estimate the width of the output interval obtained with the use of a given aggregation operator. We will concentrate on the case where all input intervals are of the same width. Preservation of the width of intervals of the same width is only one of the possible aspects of considerations. We may be interesting in determining the width of the output interval when the widths of input intervals are given (they may differ from each other). Whether or not an aggregation operator decreases, increases or preserves the width of intervals strongly depends on the concrete formula describing this aggregation operator. This is why we do not give results for the classes of aggregation operators but for some of the examples.

Proposition 2.16 (cf. [62]) *The representable interval-valued weighted mean (cf. Example 2.21, the third item) preserves the width of intervals of the same width.*

Let us recall that the representable interval-valued weighted mean is a pos-aggregation function (cf. Corollary 2.2), nec-aggregation function (cf. Theorem 2.9) and it is linear with respect to some linear orders such as $\leq_{Lex1}$, $\leq_{Lex2}$ and $\leq_{XY}$ (cf. Example 2.21).

Proposition 2.17 *Let $n \geqslant 2$. If $\mathscr{A}$ is an n-argument pseudomin or a pseudomax A_1A_2-representable aggregation function such that $A_1 = A_2$ (so it is a nec-aggregation function), A_1 is equal to the arithmetic mean and for $\boldsymbol{x}_1, \ldots, \boldsymbol{x}_n$ the* *width of intervals is respectively* $w_1 = \cdots = w_n = w$, *then the width of the output interval* $\mathscr{A}(\mathbf{x}_1, \ldots, \mathbf{x}_n)$ *equals* $\frac{n-1}{n}w$.

Proof Let $\overline{x}_i - \underline{x}_i = w$ for $i = 1, \ldots, n$. Thus if for example

$$\max\left(\frac{\underline{x}_1 + \overline{x}_2 + \cdots + \overline{x}_n}{n}, \frac{\overline{x}_1 + \underline{x}_2 + \overline{x}_3 + \cdots + \overline{x}_n}{n}, \ldots, \frac{\overline{x}_1 + \cdots + \overline{x}_{n-1} + \underline{x}_n}{n}\right) =$$

$$\frac{\underline{x}_1 + \overline{x}_2 + \cdots + \overline{x}_n}{n},$$

then

$$\overline{\mathscr{A}}(\mathbf{x}_1, \ldots, \mathbf{x}_n) - \underline{\mathscr{A}}(\mathbf{x}_1, \ldots, \mathbf{x}_n) = \frac{(\overline{x}_2 - \underline{x}_2) + (\overline{x}_3 - \underline{x}_3) + \cdots + (\overline{x}_n - \underline{x}_n)}{n} = \frac{n-1}{n}w.$$

For the remaining cases with respect to the maximum we get the same width of the output interval $\mathscr{A}(\mathbf{x}_1, \ldots, \mathbf{x}_n)$, i.e. $\frac{n-1}{n}w$.

Let us note that with the large number n of inputs the obtained width of the output interval is tending to the w value.

If the width of input intervals are not equal, then a pseudomin or a pseudomax A_1A_2-representable aggregation function (cf. (2.38), (2.37)) with $A_1 = A_2$ equal to the arithmetic mean, decreases at least one of the input widths. Let us analyze it for the case of a pseudomax representable aggregation function and $n = 2$. Let $\overline{x} - \underline{x} = w_1$, $\overline{y} - \underline{y} = w_2$. Indeed, for $\mathscr{A}([\underline{x}, \overline{x}], [\underline{y}, \overline{y}]) = \left[\frac{\underline{x}+\underline{y}}{2}, \max\left(\frac{\underline{x}+\overline{y}}{2}, \frac{\overline{x}+\underline{y}}{2}\right)\right]$, if for example $\underline{x} + \overline{y} \geqslant \overline{x} + \underline{y}$ (which means that $w_2 \geqslant w_1$), then the width of the output interval $\mathscr{A}(\mathbf{x}, \mathbf{y})$ equals $\frac{w_2}{2}$.

We will now compare, with respect to the obtained width of the output interval, representable interval-valued aggregation functions with pseudomax or pseudomin representable aggregation functions (we will consider operators that have the same component functions, cf. Definitions 2.16 and 2.19).

Proposition 2.18 *Let be given a representable interval-valued aggregation function with components* A_1, A_2 *and a pseudomax representable aggregation function (or a pseudomin representable aggregation function) with the same components* A_1, A_2 *as for the given representable interval-valued aggregation function. Thus the width of the output interval of the representable interval-valued aggregation function is greater then or equal to the width of the output interval of the pseudomax* A_1A_2*-representable aggregation function (or the pseudomin* A_1A_2*-representable aggregation function).*

Proof Let us consider the case of a pseudomax A_1A_2-representable aggregation function. Since the lower ends of the output intervals are the same we will compare the upper ends of the output intervals of both aggregation operators. Let the width of an output interval for a representable interval-valued aggregation function $\mathscr{A}$ be denoted by $\mathscr{A}_w$, then $\mathscr{A}_w(\mathbf{x}_1, \ldots, \mathbf{x}_n) = A_2(\overline{x}_1, \ldots, \overline{x}_n) - A_1(\underline{x}_1, \ldots, \underline{x}_n)$. Now, let the width of an output interval for a pseudomax representable aggregation function $\mathscr{B}$ be denoted by $\mathscr{B}_w$, then we see that

$$A_2(\underline{x}_1, \overline{x}_2, \ldots, \overline{x}_n) \leqslant A_2(\overline{x}_1, \overline{x}_2, \ldots, \overline{x}_n),$$

$$A_2(\overline{x}_1, \underline{x}_2, \overline{x}_3, \ldots, \overline{x}_n) \leqslant A_2(\overline{x}_1, \overline{x}_2, \ldots, \overline{x}_n),$$

etc. and finally

$$A_2(\overline{x}_1, \ldots, \overline{x}_{n-1}, \underline{x}_n) \leqslant A_2(\overline{x}_1, \overline{x}_2, \ldots, \overline{x}_n).$$

As a result

$$\max(A_2(\underline{x}_1, \overline{x}_2, \ldots, \overline{x}_n), A_2(\overline{x}_1, \underline{x}_2, \overline{x}_3, \ldots, \overline{x}_n), \ldots, A_2(\overline{x}_1, \ldots, \overline{x}_{n-1}, \underline{x}_n)) \leqslant$$
$$A_2(\overline{x}_1, \overline{x}_2, \ldots, \overline{x}_n).$$

This dependence holds regardless the width of input intervals (they may be equal for each input or not). As a result

$$\mathscr{A}_w(\mathbf{x}_1, \ldots, \mathbf{x}_n) \geqslant \mathscr{B}_w(\mathbf{x}_1, \ldots, \mathbf{x}_n) =$$

$$\max(A_2(\underline{x}_1, \overline{x}_2, \ldots, \overline{x}_n), A_2(\overline{x}_1, \underline{x}_2, \overline{x}_3, \ldots, \overline{x}_n), \ldots, A_2(\overline{x}_1, \ldots, \overline{x}_{n-1}, \underline{x}_n)) - A_1(\underline{x}_1, \ldots, \underline{x}_n).$$

Similar justification may be provided for the pseudomin $A_1 A_2$-representable aggregation function. This finishes the proof.

Let us consider pos-aggregation functions (2.43) and (2.44). Since by (2.19) we have

$$\frac{\overline{x}_1 + \overline{x}_2 + \cdots + \overline{x}_n}{n} \leqslant \frac{\overline{x}_1^2 + \overline{x}_2^2 + \cdots + \overline{x}_n^2}{\overline{x}_1 + \overline{x}_2 + \cdots + \overline{x}_n},$$

by Proposition 2.16 for $\mathscr{A}$ defined by (2.43) and $\overline{x}_i - \underline{x}_i = w$ for $i = 1, \ldots, n$ we obtain

$$\overline{\mathscr{A}}(\mathbf{x}_1, \ldots, \mathbf{x}_n) - \underline{\mathscr{A}}(\mathbf{x}_1, \ldots, \mathbf{x}_n) \geqslant w.$$

If for (2.43) we have

$$\overline{\mathscr{A}}(\mathbf{x}_1, \ldots, \mathbf{x}_n) - \underline{\mathscr{A}}(\mathbf{x}_1, \ldots, \mathbf{x}_n) = w_1,$$

then by (2.17) the width of the output interval for (2.44) is greater then or equal to w_1. As a result we are able to estimate the width of the output intervals for the considered aggregation operators.

However, pos-aggregation functions do not always increase the width of an output interval (in the case of equal widths of inputs). The interval-valued arithmetic mean is also a pos-aggregation function but, by Proposition 2.16, it preserves the width of input intervals of the same width.

Ii is worth to mention that in [62] it was stressed the problem of preservation of the width of intervals with the following approach. If it is assumed that the exact membership value is an element in the membership interval, then if two elements have the same interval memberships this does not mean that their corresponding real-valued membership are the same. Hence it may be natural to expect that we cannot get less uncertainty when comparing them. As a result, if we start with a given uncertainty (measured by the width of the intervals that we are considering), this uncertainty should not decrease when we make calculations, and we can only expect that it does not increase, at least, if all the initial intervals have the same width (corresponding to the same uncertainty). This may be certainly an important issue for some applications.

References

1. Marichal, J.L.: On an axiomatization of the quasi-arithmetic mean values without the symmetry axiom. Aequationes Math. **59**, 74–83 (2000)
2. Bullen, P.S., Mitrinović, D.S., Vasić, P.M.: Means and Their Inequalities. Reidel, Dordrecht (1988)
3. Ostasiewicz, S., Ostasiewicz, W.: Means and their applications. Ann. Oper. Res. **97**, 337–355 (2000)
4. Cauchy, A.L.: Cours d'analyse de l'Ecole Royale Polytechnique. Analyse Algébraique. vol. **1**, Debure, Paris (1821)
5. Chisini, O.: Sul concetto di media. Periodico di matematiche **9**(4), 106–116 (1929)
6. Finetti, B.: Sul concetto di media. Giorn. Ital. Attuari **2**(3), 369–396 (1931)
7. Ricci, U.: Confronti tra medie. Giorn. Economisti e Rivista di Statistica **26**, 38–66 (1935)
8. Kolmogorov, A.N.: Sur la notion de la moyenne. Atti Accad. Naz. Lincei Mem. Cl. Sci. Fis. Mat. Natur. Sez. **12**(6), 388–391 (1930)
9. Nagumo, M.: Über eine klasse der mittelwerte. Jpn. J. Math. **7**, 71–79 (1930)
10. Kitagawa, T.: On some class of weighted means. Proc. Phys.-Math. Soc. Jpn. **16**, 311–320 (1934)
11. Aczél, J.: Lectures on Functional Equations and Their Applications. Academic, New York (1966)
12. Aczél, J.: On mean values. Bull. Am. Math. Soc. **54**, 392–400 (1948)
13. Aczél, J., Alsina, C.: Synthesizing judgments: a functional equations approach. Math. Model. **9**, 311–320 (1987)
14. Aczél, J., Dhombres, J.: Functional Equations in Several Variables with Applications to Mathematics, Information Theory and to the Natural and Social Sciences. Cambridge University Press, Cambridge (1989)
15. Fodor, J., Marichal, J.: On nonstrict means. Aequationes Math. **54**, 308–327 (1997)
16. Hardy, G.H., Littlewood, J., Pólya, G.: Inequalities. Cambridge University Press, Cambridge (1955)
17. Bajraktarevič, M.: Sur une équation fonctionelle aux valeurs moyennes. Glasnik Mat. - Fiz. Astr. **13**, 243–248 (1958)
18. Páles, Z.: Characterization of quasideviation means. Acta Math. Acad. Sci. Hungar. **40**(3–4), 243–260 (1982)
19. Beliakov, G., Špirková, J.: Weak monotonicity of Lehmer and Gini means. Fuzzy Sets Syst. **299**, 26–40 (2016)
20. Beliakov, G., Bustince, H., Calvo, T.: A Practical Guide to Averaging Functions. Studies in fuzziness and soft computing. Springer International Publishing, Switzerland (2016)

21. Qi, F.: Generalized abstracted mean values. J. Ineq. Pure and Appl. Math. **1** (1), Art. 4 (2000) Available online at http://jipam.vu.edu.au/
22. Yager, R.R.: On ordered weighting averaging operators in multicriteria decision making. IEEE Trans. Syst. Man Cybernet. **18**, 183–190 (1988)
23. Fodor, J., Marichal, J., Roubens, M.: Characterization of the ordered weighted averaging operators. IEEE Trans. Fuzzy Syst. **3**(2), 236–240 (1995)
24. Fodor, J., Roubens, M.: Fuzzy Preference Modelling and Multicriteria Decision Support. Kluwer Academic Publishers, Dordrecht (1994)
25. Calvo, T., Kolesárová, A., Komorniková, M., Mesiar, R.: Aggregation operators: properties, classes and construction methods. In: Calvo, T., et al. (eds.) Aggregation Operators, pp. 3–104. Physica-Verlag, Heidelberg (2002)
26. Pradera, A., Beliakov, G., Bustince, H., De Baets, B.: A review of the relationships between implication, negation and aggregation functions from the point of view of material implication. Inf. Sci. **329**, 357–380 (2016)
27. Yager, R.R., Rybalov, A.: Uninorm aggregation operators. Fuzzy Sets Syst. **80**, 111–120 (1996)
28. Calvo, T., De Baets, B., Fodor, J.: The functional equations of Frank and Alsina for uninorms and nullnorms. Fuzzy Sets Syst. **120**, 385–394 (2001)
29. Bustince, H., Fernandez, J., Mesiar, R., Montero, J., Orduna, R.: Overlap functions. Nonlinear Anal.-Theor **72**, 1488–1499 (2010)
30. Maes, K.C., Saminger, S., De Baets, B.: Representation and construction of self-dual aggregation operators. Eur. J. Oper. Res. **177**, 472–487 (2007)
31. Mesiar, R., Kolesárová, A., Bustince, H., Fernandez, J.: Dualities in the class of extended Boolean functions. Fuzzy Sets Syst. **332**, 78–92 (2018)
32. Bustince, H., Fernandeza, J., Kolesarova, A., Mesiar, R.: Directional monotonicity of fusion functions. Eur. J. Oper. Res. **244**, 300–308 (2015)
33. Lucca, G., Sanz, J.A., Dimuro, G.P., Bedregal, B., Mesiar, R., Kolesárová, A., Bustince, H.: Pre-aggregation functions: construction and an application. IEEE Trans. Fuzzy Syst. **24**(2), 260–272 (2016)
34. Wilkin, T., Beliakov, G.: Weakly monotone aggregation functions. Int. J. Intell. Syst. **30**, 144–169 (2015)
35. Bustince, H., Barrenechea, E., Sesma-Sara, M., Lafuente, J., Dimuro, G. P., Mesiar, R., Kolesárová, A.: Ordered directionally monotone functions. Justification and application. IEEE Trans. Fuzzy Syst. (2017). https://doi.org/10.1109/TFUZZ.2017.2769486
36. Sesma-Sara, M.,Lafuente, J., Roldán, A., Mesiar, R., Bustince, H.: Strengthened ordered directionally monotone functions. Links between the different notions of monotonicity. Fuzzy Sets Syst. (2018). https://doi.org/10.1016/j.fss.2018.07.007
37. Tardiff, R.M.: Topologies for probabilistic metric spaces. Pac. J. Math. **65**, 233–251 (1976)
38. Schweizer, B., Sklar, A.: Probabilistic Metric Spaces. North Holland, New York (1983)
39. Saminger-Platz, S.: The dominance relation in some families of continuous Archimedean t-norms and copulas. Fuzzy Sets Syst. **160**, 2017–2031 (2009)
40. Saminger, S., Mesiar, R., Bodenhofer, U.: Domination of aggregation operators and preservation of transitivity. Int. J. Uncertain. Fuzz. Knowledge-Based Syst. **10**/s, 11–35 (2002)
41. Bentkowska, U., Drewniak, J., Drygaś, P., Król, A., Rak, E.: Dominance of binary operations on posets. In: Atanassov, K.T., et al. (eds.) Uncertainty and Imprecision in Decision Making and Decision Support: Cross-Fertilization, New Models and Applications. IWIFSGN 2016. Advances in Intelligent Systems and Computing, vol. 559, pp. 143–152. Springer, Cham (2018)
42. Saminger-Platz, S., Mesiar, R., Dubois, D.: Aggregation operators and commuting. IEEE Trans. Fuzzy Syst. **15**(6), 1032–1045 (2007)
43. Bentkowska, U., Król, A.: Preservation of fuzzy relation properties based on fuzzy conjunctions and disjunctions during aggregation process. Fuzzy Sets Syst. **291**, 98–113 (2016)
44. Drewniak, J., Dudziak, U.: Preservation of properties of fuzzy relations during aggregation processes. Kybernetika **43**, 115–132 (2007)
45. Dubois, D., Prade, H.: Weighted minimum and maximum operations in fuzzy set theory. Inf. Sci. **39**, 205–210 (1986)

46. Marrara, S., Pasi, G., Viviani, M.: Aggregation operators in information retrieval. Fuzzy Sets Syst. **324**, 3–19 (2017)
47. Lucca, G., Sanz, J.A., Dimuro, G.P., Bedregal, B., Asiain, M.J., Elkano, M., Bustince, H.: CC-integrals: choquet-like copula-based aggregation functions and its application in fuzzy rule-based classification systems. Knowledge-Based Syst. **119**, 32–43 (2017)
48. Dimuro, G.P., Lucca, G., Sanz, J.A., Bustince, H., Bedregal, B.: CMin-Integral: a choquet-like aggregation function based on the minimum t-norm for applications to fuzzy rule-based classification systems. In: Torra, V., et al. (eds.) Aggregation Functions in Theory and in Practice. Advances in Intelligent Systems and Computing, vol. 581, pp. 83–95. Springer, Cham (2018)
49. Beliakov, G., Pradera, A., Calvo, T.: Aggregation Functions: A Guide for Practitioners. Springer, Berlin (2007)
50. Grabisch, M., Marichal, J.-L., Mesiar, R., Pap, E.: Aggregation Functions. Cambridge University Press, Cambridge (2009)
51. Torra, V., Narukawa, Y.: Modeling Decisions: Information Fusion and Aggregation Operators. Springer, Berlin (2007)
52. Dyczkowski, K.: Intelligent medical decision support system based on imperfect information. The Case of Ovarian Tumor Diagnosis. Studies in Computational Intelligence. Springer, Cham (2018)
53. Bentkowska, U., Pękala, B.: Diverse classes of interval-valued aggregation functions in medical diagnosis support. In: Medina, J. et al. (eds.) IPMU 2018, pp. 391–403. Springer International Publishing AG, part of Springer, CCIS 855 (2018)
54. Komorníková, M., Mesiar, R.: Aggregation functions on bounded partially ordered sets and their classification. Fuzzy Sets Syst. **175**, 48–56 (2011)
55. Deschrijver, G.: Arithmetic operators in interval-valued fuzzy set theory. Inf. Sci. **177**, 2906–2924 (2007)
56. Drygaś, P., Pękala, B.: Properties of decomposable operations on some extension of the fuzzy set theory. In: Atanassov, K.T., Hryniewicz, O., Kacprzyk, J. et al., (eds.) Advances in Fuzzy Sets, Intuitionistic Fuzzy Sets. Generalized Nets and Related Topics, pp. 105–118. EXIT, Warsaw (2008)
57. Deschrijver, G.: Quasi-arithmetic means and OWA functions in interval-valued and Atanassovs intuitionistic fuzzy set theory. In: Galichet, S. et al. (eds.) Proceedings of EUSFLAT-LFA 2011, 18-22.07.2011, Aix-les-Bains, pp. 506–513. France (2011)
58. Bentkowska, U., Pękala, B., Bustince, H., Fernandez, J., Jurio, A., Balicki, J.: N-reciprocity property for interval-valued fuzzy relations with an application to group decision making problems in social networks. International Journal of Uncertainty, Fuzziness and Knowledge-Based Systems **25**(Suppl. 1), 43–72 (2017)
59. Deschrijver, G., Cornelis, C., Kerre, E.: On the representation of intuitonistic fuzzy t-norms and t-conorms. IEEE Trans. Fuzzy Syst. **12**, 45–61 (2004)
60. Deschrijver, G.: Generators of t-norms in interval-valued fuzzy set theory. In: Montseny, E. and Sobrevilla, P. (eds.) Proceedings of the Joint 4th Conference of the European Society for Fuzzy Logic and Technology and the 11th Rencontres Francophones sur la Logique Floue et ses Applications, Barcelona, Spain, September 7–9, 2005, pp. 253–258. Spain (2005)
61. Bentkowska, U.: New types of aggregation functions for interval-valued fuzzy setting and preservation of pos-B and nec-B-transitivity in decision making problems. Inf. Sci. **424**, 385–399 (2018)
62. Zapata, H., Bustince, H., Montes, S., Bedregal, B., Dimuro, G.P., Takáč, Z., Baczyński, M., Fernandez, J.: Interval-valued implications and interval-valued strong equality index with admissible orders. Int. J. Approx. Reason. **88**, 91–109 (2017)
63. Bustince, H., Galar, M., Bedregal, B., Kolesárová, A., Mesiar, R.: A new approach to interval-valued Choquet integrals and the problem of ordering in interval-valued fuzzy sets applications. IEEE Trans. Fuzzy Syst. **21**(6), 1150–1162 (2013)
64. Bedregal, B., Bustince, H., Dimuro, G.P., Fernandez, J.: Generalized interval-valued OWA operators with interval weights derived from interval-valued overlap functions. Int. J. Approx. Reason. **90**, 1–16 (2017)

65. Qiao, J., Hu, B.Q.: On interval additive generators of interval overlap functions and interval grouping functions. Fuzzy Sets Syst. **323**, 19–55 (2017)
66. Bustince, H., Fernandez, J., Mesiar, R., Montero, J., Orduna, R.: Overlap index, overlap functions and migrativity. In: Carvalho, J. P., Dubois, D., Kaymak, U. and Sousa, J. M. C. (eds.) Proceedings of the Joint 2009 International Fuzzy Systems Association World Congress and 2009 European Society of Fuzzy Logic and Technology Conference, Lisbon, Portugal, July 20–24, pp. 300–305. ISBN: 978 – 989 – 95079 – 6 – 8 (2009)
67. Bustince, H., Pagola, M., Mesiar, R., Hüllermeier, E., Herrera, F.: Grouping, overlaps, and generalized bientropic functions for fuzzy modeling of pairwise comparisons. IEEE Trans. Fuzzy Syst. **20**, 405–415 (2012)
68. Dimuro, G.P., Bedregal, B.C., Santiago, R.H.N., Reiser, R.H.S.: Interval additive generators of interval t-norms and interval t-conorms. Inf. Sci. **181**, 3898–3916 (2011)
69. Pękala, B.: Uncertainty data in interval-valued fuzzy set theory. Properties, Algorithms and Applications. Studies in Fuzziness and Soft Computing. Springer, Cham (2019)
70. Fuchs, L.: Partially Ordered Algebraic Systems. Pergamon Press, Oxford (1963)
71. Elkano, M., Sanz, J.A., Galar, M., Pękala, B., Bentkowska, U., Bustince, H.: Composition of interval-valued fuzzy relations using aggregation functions. Inf. Sci. **369**, 690–703 (2016)

Part II
Applications

This part of the book presents application areas of the interval-valued methods, especially interval-valued aggregation operators that were discussed in the first part of the book. There are considered decision making and classification problems. Firstly, the preservation of interval-valued fuzzy relation properties in aggregation process is studied. There are considered the pos-B-transitivity and nec-B-transitivity properties as the ones which are very important for preference relations in multicriteria decision making. The overwhelming majority of this part of the book is devoted to classification problems. There are presented methods of optimization for the k-NN algorithm considered on the data sets with large number of missing values or large number of attributes. The case of large number of attributes is presented in connection with DNA microarray methods. In both experiments there are compared methods of aggregation with the use of aggregation on the real-line and intervalvalued aggregation methods. The analysis of the results of both performed experiments proves the superiority of interval-valued methods and efficacy of the recently introduced pos-aggregation functions and nec-aggregation functions. There is also provided the chapter presenting the behavior of some new classes of aggregation operators in one of the existing decision support systems.

Chapter 3
Decision Making Using Interval-Valued Aggregation

The speed of decision making is the essence of good governance.
Piyush Goyal

In this chapter we present the results connected with multicriteria (or similarly multi-agent) decision making problems under uncertainty. The aspects of decision making form a wide branch of considerations both in fuzzy sets theory and its extensions (cf. [1–3]). We consider only some of the basic concepts for the case of interval-valued fuzzy relations. Here, there is examined preservation of transitivity properties by aggregation operators. Similar considerations may be performed for the remaining properties. We concentrate on transitivity as an exemplary property since this is one of the most important properties which may guarantee consistency of choices of decision makers. Namely, following crisp notion of transitivity we see that if x is preferred to y and y is preferred to z, mathematically $R(x, y)$ and $R(y, z)$, then x should be preferred to z, in mathematical notions it means that $R(x, z)$ holds. This is intuitive and natural assumption. We provide the results connected with the notions of pos-B-transitivity, nec-B-transitivity and preservation of these properties by aggregation operators in decision making. We propose to apply the respective notion of transitivity and aggregation method, depending on the requirements of the given problem. Namely, if for a given problem we require that at least one element in the first interval is smaller or equal to at least one element in the second interval, then the notions related to $\preceq_\pi$ would be suitable (pos-B-transitivity, pos-aggregation function). If for a given problem we require that each element in the first interval is smaller or equal to each element in the second interval, then the notions related to $\preceq_\nu$ would be suitable (nec-B-transitivity, nec-aggregation function). We think that such approach may lead to the more meaningful results and better choice of the solution alternatives (however, please note that the mentioned classes of aggregation operators are not disjoint, cf. Corollary 2.2 and Theorem 2.9).

U. Bentkowska, *Interval-Valued Methods in Classifications and Decisions*,
Studies in Fuzziness and Soft Computing 378,
https://doi.org/10.1007/978-3-030-12927-9_3

In multicriteria decision making a decision maker has to choose among the alternatives with respect to a set of criteria. Let $card\ X = m, m \in \mathbb{N}, X = \{x_1, \ldots, x_m\}$ be a set of alternatives, $K = \{k_1, \ldots, k_n\}$ be a set of criteria on the base of which the alternatives are evaluated. $R_1, \ldots, R_n$ be interval-valued fuzzy relations on a set X corresponding to each criterion represented by matrices, i.e. $R_k \in \mathscr{IVFR}(X)$, $k = 1, \ldots, n$, $n \in \mathbb{N}$. Similarly, if we consider multiagent decision making problems, relations $R_1, \ldots, R_n$ represent the preferences of each agent (certainly, we can combine these two situations, i.e. many criteria and many agents). Relation $R_{\mathscr{F}} = \mathscr{F}(R_1, \ldots, R_n)$ is supposed to help the decision maker to make up his/her mind. Some functions $\mathscr{F}$ may be more adequate for aggregation than the others since they may (or not) preserve the required properties of relations $R_1, \ldots, R_n \in \mathscr{IVFR}(X)$. This is why preservation of these properties may be an interesting issue and required in aggregation process for multicriteria or multiagent decision making problems. Other properties of interval-valued fuzzy relations also may provide the interpretation of choices of decision makers. This is why it is important to check the connections between input fuzzy relations and the output one. From practical point of view, for multicriteria or multiagent decision making problems, it means that we check if the particular choices of decision makers are of the same type (or they are consistent) as the aggregated fuzzy relation presents.

The comparability relations $\preceq_\pi$ and $\preceq_\nu$, and as a consequence adequate type of aggregation functions and transitivity properties (which involve in their notions the mentioned comparability relations), enable to use them if in a decision making problem there is required one of two interpretations, namely *possible* or *necessary*.

3.1 Preservation of Interval-Valued Fuzzy Relation Properties in Aggregation Process

In this section we consider the aggregation of interval-valued fuzzy relations having pos–B–transitivity and nec–B–transitivity property. To simplify the notions, the results are presented for aggregation of two input fuzzy relations. Certainly, the results may be transformed for n-argument case of the input fuzzy relations. The problem of preservation of transitivity properties is considered, i.e. if we assume that $R_1, R_2 \in \mathscr{IVFR}(X)$ have some property, then $R_{\mathscr{F}} = \mathscr{F}(R_1, R_2)$ also has the same property. Although $\mathscr{F} : (L^I)^2 \to L^I$ may be an aggregation operator (of some type) we consider $\mathscr{F}$ with no pre-assumed properties. This approach enables to get more general results when describing operator $\mathscr{F}$. We generally intend to consider the same type of property and aggregation operator, namely based on the same type of comparability relation $\preceq_\pi$ or $\preceq_\nu$. However, to complete the results we also consider the mixture of an aggregation type and the type of property. We examine subfamilies of aggregation operators which have some representation, i.e. representable or pseudo-representable ones.

The next statements will be presented for general case of decomposable operators and then appropriate conclusions for considered here families of aggregation operators will be provided.

Theorem 3.1 ([4]) *Let $\mathscr{F}$ be decomposable and F_1 be increasing, $F_1 \gg B$, $F_1 \leqslant F_2$. If $R_1, R_2 \in \mathscr{IVFR}(X)$ are pos–B–transitive, then $R_{\mathscr{F}}(x, y) = \mathscr{F}(R_1(x, y), R_2(x, y))$ is pos–B–transitive.*

Corollary 3.1 *Let $\mathscr{F}$ be a decomposable interval-valued aggregation function (decomposable pos-aggregation function, or decomposable nec-aggregation function, then $F_1 = F_2$ according to Theorem 2.9) and $F_1 \gg B$. If $R_1, R_2 \in \mathscr{IVFR}(X)$ are pos–B–transitive, then $R_{\mathscr{F}}(x, y) = \mathscr{F}(R_1(x, y), R_2(x, y))$ is pos–B–transitive.*

Let us notice that dominance over operation B is not necessary for preservation of pos–B–transitivity.

Remark 3.1 Most of the pos-aggregation functions from Examples 2.28 and 2.29 preserve pos–B–transitivity. For operations (2.59)–(2.64) in Example 2.29 it is enough to assume that B is increasing. For (2.49), (2.50), (2.55) in Example 2.28, under assumption that for input relations $R_1, R_2 \in \mathscr{IVFR}(X)$ we have $R_1(x, y) \neq [1, 1]$, $R_2(x, y) \neq [1, 1]$ for any $x, y \in X$, it is enough to assume that $B(0, 0) = 0$. For (2.51), (2.52), (2.56) in Example 2.28, under assumption that for input relations $R_1, R_2 \in \mathscr{IVFR}(X)$ we have $R_1(x, y) \neq [0, 0]$, $R_2(x, y) \neq [0, 0]$ for any $x, y \in X$, we get straightforward (by definition of aggregation functions) preservation of pos–B–transitivity.

Theorem 3.2 ([4]) *Let $\mathscr{F}$ be decomposable and F_1 be increasing, $F_1 \gg B$, $F_1 = F_2$. If $R_1, R_2 \in \mathscr{IVFR}(X)$ are nec–B–transitive, then $R_{\mathscr{F}}(x, y) = \mathscr{F}(R_1(x, y), R_2(x, y))$ is nec–B–transitive.*

Corollary 3.2 *Let $\mathscr{F}$ be a decomposable interval-valued aggregation function such that $F_1 = F_2$ (decomposable pos-aggregation function such that $F_1 = F_2$ or decomposable nec-aggregation function, then $F_1 = F_2$ according to Theorem 2.9), and $F_1 \gg B$. If $R_1, R_2 \in \mathscr{IVFR}(X)$ are nec–B–transitive, then $R_{\mathscr{F}}(x, y) = \mathscr{F}(R_1(x, y), R_2(x, y))$ is nec–B–transitive.*

Remark 3.2 Let us note that decomposability is not necessary for preservation of nec–B–transitivity property. The following interval-valued aggregation function $\mathscr{A} : (L^I)^2 \to L^I$, $\mathscr{A}(\mathbf{x}, \mathbf{y}) = \left[\frac{\underline{y}+\frac{\underline{x}+\overline{x}}{2}}{2}, \frac{\overline{x}\overline{y}}{2}\right]$ is not decomposable, it belongs to $\mathscr{A}_\nu$ (cf. Example 2.40) and it preserves nec–B–transitivity for any $B : [0, 1]^2 \to [0, 1]$ such that B is dominated by the arithmetic mean M, i.e. $M \gg B$ [4].

Now we turn to the problem of preservation of pos-B-transitivity and nec-B-transitivity properties by other than decomposable operators. We focus on pseudomin and pseudomax aggregation functions.

Theorem 3.3 ([4]) *Let $\mathscr{A}$ be a pseudomin A_1A_2-representable aggregation function and $A_1 = A_2$ (by Proposition 2.7 it means that $\mathscr{A} \in \mathscr{A}_\nu$). If $A_1 \gg B$, then $\mathscr{A}$ preserves nec–B–transitivity.*

Theorem 3.4 ([4]) *Let $\mathscr{A}$ be a pseudomax A_1A_2-representable aggregation function and $A_1 = A_2$ (by Proposition 2.6 it means that $\mathscr{A} \in \mathscr{A}_\nu$). If $A_1 \gg B$, B is increasing, then $\mathscr{A}$ preserves nec–B–transitivity.*

Below we present the results connected with preservation of pos–B–transitivity by pseudomin and pseudomax aggregation functions. Sufficient conditions for preservation of pos–B–transitivity in these cases are rather strong (but possible to be fulfilled), since they are connected with dominance of an aggregation function over minimum and dominance of maximum over an aggregation function (cf. Theorems 2.5 and 2.6).

Theorem 3.5 ([4]) *Let $\mathscr{A}$ be a pseudomax A_1A_2-representable aggregation function. If $A_1 \gg B$, $\max \gg A_2$, then $\mathscr{A}$ preserves pos–B–transitivity.*

Theorem 3.6 ([4]) *Let $\mathscr{A}$ be a pseudomin A_1A_2-representable aggregation function. If $A_1 \gg B$, $A_1 \gg \min$, B is increasing, then $\mathscr{A}$ preserves pos–B–transitivity.*

There can be given many examples of aggregation operators (diverse types presented here) preserving pos–B–transitivity and nec–B–transitivity. Some of them follow from Example 2.9 and Corollary 2.1. Below we give only one of the simplest examples which covers all presented above theorems for the considered here transitivity properties.

Example 3.1 Since minimum dominates any fuzzy conjunction C (cf. Corollary 2.1), an interval-valued aggregation function $\mathscr{A}(\mathbf{x}, \mathbf{y}) = [\min(\underline{x}, \underline{y}), \min(\overline{x}, \overline{y})]$, where also $\mathscr{A} \in \mathscr{A}_\nu$ and $\mathscr{A} \in \mathscr{A}_\pi$, preserves pos–$C$–transitivity and nec–$C$–transitivity. Similarly,

$$\mathscr{A}(\mathbf{x}, \mathbf{y}) = [\min(\underline{x}, \underline{y}), \max(\min(\underline{x}, \overline{y}), \min(\overline{x}, \underline{y}))]$$

is a pseudomax A_1A_2-representable aggregation function that preserves nec–C–transitivity.

Aggregation operator $\mathscr{A}(\mathbf{x}, \mathbf{y}) = [\min(A(\underline{x}, \overline{y}), A(\overline{x}, \underline{y})), A(\overline{x}, \overline{y})]$ is a pseudomin A_1A_2-representable aggregation function, where $A_1 = A_2 = A$ is a weighted minimum, and it preserves nec–C–transitivity and pos–C–transitivity where C is an arbitrary t-seminorm (cf. Corollary 2.1, Theorem 2.5).

In the next section the presented here results will be applied in a given algorithm for decision making.

3.2 Multicriteria Decision Making Algorithm

We present here one of the possible algorithms to obtain the final solution from a given set of alternatives. First, we recall the suitable notions and theoretical results that are applied in the algorithm. For decision making problems (at the stage of choosing the best alternative) it is often required that the obtained relation is a reciprocal one. We recall this notion in the originally considered version, i.e. with the standard negation $N(x) = 1 - x$ applied (cf. (3.1)). The reciprocity property with respect to other negations (called N-reciprocity) along with the use in decision making problems, was considered in [5] (in the case of N-reciprocity in (3.1) instead of $N(x) = 1 - x$, an arbitrary strong fuzzy negation N is used). We provide here suitable concepts and results involving interval-valued fuzzy relations and present them in the matrix-notion type (cf. Remark 1.1). This is due to the application reasons, since decision making problems are considered on a finite set.

Definition 3.1 ([6]) An interval-valued fuzzy reciprocal relation R on a set X is a matrix $R = (R_{ij})_{m\times m}$ with $R_{ij} = [\underline{R}(i, j), \overline{R}(i, j)]$, for all $i, j \in \{1, \ldots, m\}$ where

$$R_{ij} \in L^I,\ R_{ji} = 1 - R_{ij} = [1 - \overline{R}(i, j), 1 - \underline{R}(i, j)],\ R_{ii} = [0.5, 0.5]. \tag{3.1}$$

The interval R_{ij} indicates the interval-valued preference degree or intensity of the alternative x_i over alternative x_j and $\underline{R}(i, j)$, $\overline{R}(i, j)$ are the lower and upper limits of r_{ij}, respectively. Moreover,

- $R_{ij} = [0.5, 0.5]$ indicates indifference between x_i and x_j,
- $R_{ij} > [0.5, 0.5]$ indicates that x_i is strictly preferred to x_j, and
- $R_{ij} = [1, 1]$ indicates that x_i is absolutely preferred to x_j.

The assumption $R_{ji} = 1 - R_{ij}$ for $i, j \in \{1, \ldots, n\}$, which follows from the reciprocity property, is rather strong and frequently violated by decision makers in real-life situations. This is why often the normalization of an interval-valued fuzzy relation is performed to obtain an N-reciprocal relation. To normalize $R \in \mathcal{IVFR}(X)$ the following formula may be used (cf. [5], note that if R is N-reciprocal, then $R^* = R$)

$$R^*_{ij} = \begin{cases} \mathcal{N}(R_{ji}) & \text{if } R_{ij} >_{L^I} R_{ji}, \\ R_{ij} & \text{else}, \end{cases} \tag{3.2}$$

where $R_{ij} >_{L^I} R_{ji}$ means that $R_{ij} \geq_{L^I} R_{ji}$ but $R_{ij} \neq R_{ji}$ and $\mathcal{N}$ is a representable interval-valued fuzzy negation. Instead of $\leq_{L^I}$ we may also use $\preceq_\pi$ or $\preceq_\nu$.

Moreover, in decision making problems the original relation $R \in \mathcal{IVFR}(X)$ is displayed into corresponding interval-valued strict preference (P), interval-valued indifference (I) and interval-valued incomparability (J) (cf. [7–9]). Such preference structures were considered for example in [10], where representable interval-valued t-norms were used to build P, I and J. In [5] there was presented a generalized method of building the triplet (P, I, J) involving interval-valued aggregation functions and an interval-valued fuzzy negation $\mathcal{N}$ given in Definition 1.16. We may extend these notions using the concepts of pos- and nec-aggregation functions. Let $p, i, j : (L^I)^2 \to L^I$ be functions of the following form

$$P(x, y) = p(R(x, y), R(y, x)),\ x, y \in X$$

$$I(x,y) = i(R(x,y), R(y,x)),\ x,y \in X$$

$$J(x,y) = j(R(x,y), R(y,x)),\ x,y \in X.$$

Such functions do exist by invoking the axiom of Independence of Irrelevant Alternatives (as it was shown for the fuzzy case in [7] and applied also in interval-valued fuzzy case in [5]), which states that for any two alternatives x, y the values $P(x, y)$, $I(x, y)$ and $J(x, y)$ depend only on the values $R(x, y)$ and $R(y, x)$. According to the axiom of Positive Association principle, $p(x, y)$ should be nondecreasing with respect to its first argument but nonincreasing with respect to its second argument, $i(x, y)$ nondecreasing with respect to both arguments, and $j(x, y)$ nonincreasing with respect to both arguments. We will consider here monotonicity with respect to relations (1.8), (1.11) and (1.12). According to the axiom of Symmetry it should hold $i(x, y) = i(y, x)$ and $j(x, y) = j(y, x)$. As a result we obtain the following representations of P, I and J (cf. [5]). Let $\mathscr{A}$ and $\mathscr{B}$ be decomposable interval-valued aggregation functions (decomposable pos-aggregation functions or decomposable nec-aggregation functions, respectively). We consider the following interval-valued fuzzy relations:
interval-valued strict preference

$$\begin{aligned} P_{ij} &= \mathscr{A}(R_{ij}, \mathscr{N}(R_{ji})) \\ &= \mathscr{A}([\underline{R}_{ij}, \overline{R}_{ij}], [N(\overline{R}_{ji}), N(\underline{R}_{ji})]) \\ &= [A_1(\underline{R}_{ij}, N(\overline{R}_{ji})), A_2(\overline{R}_{ji}, N(\underline{R}_{ji}))], \end{aligned} \tag{3.3}$$

interval-valued indifference

$$\begin{aligned} I_{ij} &= \mathscr{B}(R_{ij}, R_{ji}) \\ &= \mathscr{B}([\underline{R}_{ij}, \overline{R}_{ij}], [\underline{R}_{ji}, \overline{R}_{ji}]) \\ &= [B_1(\underline{R}_{ij}, \underline{R}_{ji}), B_2(\overline{R}_{ij}, \overline{R}_{ji})], \end{aligned} \tag{3.4}$$

interval-valued incomparability

$$\begin{aligned} J_{ij} &= \mathscr{B}(\mathscr{N}(R_{ij}), \mathscr{N}(R_{ji})) \\ &= \mathscr{B}(\mathscr{N}([\underline{R}_{ij}, \overline{R}_{ij}]), \mathscr{N}([\underline{R}_{ji}, \overline{R}_{ji}])) \\ &= [B_1(N(\overline{R}_{ij}), N(\overline{R}_{ji})), B_2(N(\underline{R}_{ij}), N(\underline{R}_{ji}))] \end{aligned} \tag{3.5}$$

for all $i, j \in \{1, \ldots, m\}$.

Let us note that due to the standard practical considerations, these three mappings p, i, j were often assumed to be continuous (cf. [8]). Since, we propose to use decomposable operators we suggest to consider continuous components of $\mathscr{A}$ and $\mathscr{B}$. The following result extends the one presented in [5] (cf. [11]) for interval-valued aggregation functions. The result presented below involves also pos- and nec-aggregation functions.

Theorem 3.7 *Let $R \in \mathscr{IVFR}(X)$ be N-reciprocal and let P_{ij} be its associated interval-valued strict fuzzy preference relation built by (3.3). Thus $P_{ij} = R_{ij}$ for all $i, j \in \{1, \dots, m\}$ if and only if a decomposable interval-valued aggregation function $\mathscr{A}$ (decomposable pos-aggregation function or decomposable nec-aggregation function $\mathscr{A}$) is idempotent.*

Proof Let R be N-reciprocal and $\mathscr{A}([\mathbf{x}, \mathbf{y}]) = [A_1(\underline{x}, \underline{y}), A_2(\overline{x}, \overline{y})]$ be an idempotent decomposable interval-valued aggregation function (pos-aggregation function or nec-aggregation function). From idempotency of $\mathscr{A}$ we have idempotency of A_1 and A_2 (cf. Theorem 2.10). Then

$$P_{ij} = [A_1(\underline{R}_{ij}, N(\overline{R}_{ji})), A_2(\overline{R}_{ij}, N(\underline{R}_{ji}))] = [A_1(\underline{R}_{ij}, \underline{R}_{ij}), A_2(\overline{R}_{ij}, \overline{R}_{ij})] = [\underline{R}_{ij}, \overline{R}_{ij}] = R_{ij}.$$

If we assume that $P_{ij} = R_{ij}$, then analyzing the above formula we see that $\mathscr{A}$ is idempotent.

Let us note that decomposability of a nec-aggregation $\mathscr{A}$ (considered in Theorem 3.7) means that $A_1 = A_2$ and a nec-aggregation $\mathscr{A}$ is an interval-valued aggregation function (cf. Theorem 2.9). However, there are decomposable pos-aggregation functions which are not interval-valued aggregation functions (cf. Example 2.29).

There exist diverse methods that allow to find the best alternative (cf. [5, 10, 12]). These methods may be based on the voting methods and the non-dominance one originally considered for fuzzy settings (cf. [13–15]). We will adapt to our example the non-dominance method.

When we are working with numbers, a linear order is at our disposal, and we are always able to say which number is the greatest one. In order to work with intervals, if we want to compare them, the easiest way is to pick up a linear order between intervals. The methods of building a linear order between intervals were presented in Chap. 1. We may also use the comparability relations $\preceq_\pi$ or $\preceq_\nu$. However, in such cases we may need to use some additional methods to compare intervals (e.g. calculating the width of intervals). These types of methods, involving $\preceq_\pi$ or $\preceq_\nu$, are applied also in Sect. 6.2. In the present considerations to compare intervals we will apply both relation $\preceq_\pi$ (or $\preceq_\nu$) and linear orders. It is clear that the use of the non-dominance method does not guarantee that we can select a unique solution alternative. However, if we want to use linear orders, one may apply the non-dominance method several times with a different linear order each of time. In this way, we may take as solution the alternative which appears most often in the first place (in the different solutions that are obtained [12]). By Dirichlet's Box Principle, it is enough to take as many linear orders as alternatives we have plus one. In this way, we are certain that at least one of the alternatives appears twice in the first position. Let us note that in [16] there was proposed an algorithm for consensus between diverse linear orders and Choquet integrals that are used in a given problem. It is clear that diverse orders may lead to different solution alternatives. Sometimes the application determines the order to be used. There are also cases when the experts should be taken into account when choosing the order. For example, if the experts are considered to be optimistic, then it may be intuitive to use the order $\leq_{Lex2}$. On the contrary, if they are considered to be pessimistic, the order $\leq_{Lex1}$ may be more suitable. However, in many cases, we do not have this information.

Algorithm 1: Algorithm for decision making

Input:
1. set of alternatives $X = \{x_1, \ldots, x_m\}$,
2. collection $R_1, \ldots, R_n \in \mathcal{IVFR}(X)$ of interval-valued fuzzy relations,
3. $\mathbf{B} = \{B | B : [0, 1]^2 \to [0, 1]\}$ - finite set of given comparable operations including constant operations $B \equiv 0$ (denoted by B^0), $B \equiv 1$ (denoted by B^1),
4. $\mathbf{F}$ a finite set of operators $\mathcal{F}$ that preserve pos-B-transitivity (or nec–B–transitivity), where $\mathcal{F} \in \mathcal{A}_\pi$ ($\mathcal{F} \in \mathcal{A}_\nu$)
5. Interval-valued fuzzy negation $\mathcal{N}$, aggregation operator $\mathcal{A} \in \mathcal{A}_\pi$ ($\mathcal{A} \in \mathcal{A}_\nu$ or $\mathcal{A}$ - an interval-valued aggregation function) used to built P.

Output: A solution alternative from the set X.

```
begin
    for i := 1 to n do
        Check the type of pos–B–transitivity (or nec–B–transitivity).
    end
    Fix the common type of pos–B–transitivity (or nec–B–transitivity) of all R_1, ..., R_n.
    Determine the relation R_F with the use of F.
    Obtain a normalized fuzzy preference relation R*_F using (3.2).
    Compute the fuzzy strict preference relation P on the base of R*_F using Eq. (3.3).
    Build the nondominance interval-valued fuzzy set on the base of P:
```

$$ND_{IV} = \{(x_j, ND_{IV}(x_j) = [\bigvee_{i=1}^{m} (\underline{P}_{ij}), \bigvee_{i=1}^{m} (\overline{P}_{ij})]) | x_j \in X\}.$$

```
    Apply N_s, generated by the standard fuzzy negation, to the set ND_IV:
```

$$\mathcal{N}_s(ND_{IV})(x_j) = [1 - \bigvee_{i=1}^{m} (\overline{P}_{ij}), 1 - \bigvee_{i=1}^{m} (\underline{P}_{ij})].$$

```
    Order the alternatives in a non-increasing way using diverse orders.
    Find the solution alternative from the set X.
end
```

Now, a numerical example will be provided to illustrate the proposed algorithm. We present an example for preservation of pos–B–transitivity. Analogously, one may consider this algorithm for nec-transitivity involved. It is worth to mention that for decision making problems if the interval-valued fuzzy relations are considered, along with the meaning of membership grades, crucial assumption is card $X \geqslant 4$. Practically, an interval-valued fuzzy relation makes sense only if it is meaningful to compare $R(x, y)$ to $R(z, w)$ for 4-tuples of acts (x, y, z, w), that is, in the scope of preference modelling, to decide whether x is preferred (or not) to y in the same way or not as z is preferred to w (cf. [17]).

Let $X = \{x_1, x_2, x_3, x_4\}$ and be given interval-valued fuzzy relations $R_1, R_2 \in \mathcal{IVFR}(X)$ which present preferences over alternatives x_1, x_2, x_3, x_4, where

$$R_1 = \begin{bmatrix} [0.5, 0.5] & [0.2, 0.4] & [0.3, 0.4] & [0.5, 0.8] \\ [1, 1] & [0.5, 0.5] & [0.3, 0.5] & [0.5, 0.6] \\ [0.7, 0.8] & [0.6, 0.8] & [0.5, 0.5] & [0.6, 0.9] \\ [0.6, 0.7] & [0.6, 0.8] & [0.5, 0.6] & [0.5, 0.5] \end{bmatrix},$$

$$R_2 = \begin{bmatrix} [0.5, 0.5] & [0.3, 0.6] & [0.3, 0.5] & [0.8, 0.9] \\ [0.7, 0.9] & [0.5, 0.5] & [0.3, 0.4] & [0.8, 1] \\ [0, 0.1] & [0, 0.2] & [0.5, 0.5] & [1, 1] \\ [0, 0.3] & [0, 0.1] & [0, 0.2] & [0.5, 0.5] \end{bmatrix}$$

and $\mathbf{B} = \{B^0, T_D, T_L, T_P, \min, B^1\}$, where $B^0 \leqslant T_D \leqslant T_L \leqslant T_P \leqslant \min \leqslant B^1$. In the provided example, the set $\mathbf{F}$ may for example include the family of representable weighted means (cf. Theorem 3.1 and Eq. (2.33) for respective weights) that preserve pos-B-transitivity for $B \in \{B^0, T_D, T_L\}$. We see that R_1 is pos-T_L-transitive ($\underline{R}_1 \circ_{T_L} \underline{R}_1 \leqslant \overline{R}_1$, cf. Proposition 1.6) so it is also pos-T_D-transitive and pos-B^0-transitive but R_1 is not pos-B-transitive for $B \in \{T_P, \min, B^1\}$. Whereas R_2 is pos-min-transitive so it is also pos-B-transitive for $B \in \{B^0, T_D, T_L, T_P\}$. As a result both R_1 and R_2 are pos-T_L-transitive (cf. Proposition 1.8). Interval valued fuzzy relations R_1, R_2 may be aggregated for example with the use of $\mathscr{F}(\mathbf{x}, \mathbf{y}) = [0.6\underline{x} + 0.4\underline{y}, 0.6\overline{x} + 0.4\overline{y}]$ (which is a decomposable pos-aggregation function preserving pos–T_L–transitivity, cf. Theorem 3.1). As a result, $R_{\mathscr{F}} = \mathscr{F}(R_1, R_2)$ is also pos-T_L-transitive, where

$$R_{\mathscr{F}} = \begin{bmatrix} [0.5, 0.5] & [0.24, 0.48] & [0.3, 0.44] & [0.62, 0.84] \\ [0.68, 0.96] & [0.5, 0.5] & [0.3, 0.46] & [0.62, 0.76] \\ [0.42, 0.52] & [0.36, 0.56] & [0.5, 0.5] & [0.76, 0.94] \\ [0.56, 0.62] & [0.36, 0.52] & [0.3, 0.44] & [0.5, 0.5] \end{bmatrix}.$$

In the algorithm we use the dependence presented in Proposition 1.8. Since we assume that B^0 belongs to the class $\mathbf{B}$, we always may determine the common type of pos–B–transitivity (or nec–B–transitivity). For B^0 these transitivity conditions are fulfilled trivially, so the algorithm may be run to the end (however not giving an interesting result). Next, by Algorithm 1, the normalized $R^*_{\mathscr{F}}$ is obtained by (3.2), here with the use of $N(x) = 1 - x$ and $\leq_{Lex1}$, so (3.2) takes the form

$$R^*_{ij} = \begin{cases} [1 - \overline{R}_{ji}, 1 - \underline{R}_{ji}] & \text{if } R_{ij} >_{Lex1} R_{ji}, \\ R_{ij} & \text{else.} \end{cases} \tag{3.6}$$

As a result

$$R^*_{\mathscr{F}} = \begin{bmatrix} [0.5, 0.5] & [0.24, 0.48] & [0.3, 0.44] & [0.38, 0.44] \\ [0.52, 0.76] & [0.5, 0.5] & [0.3, 0.46] & [0.48, 0.64] \\ [0.56, 0.7] & [0.54, 0.7] & [0.5, 0.5] & [0.56, 0.7] \\ [0.56, 0.62] & [0.36, 0.52] & [0.3, 0.44] & [0.5, 0.5] \end{bmatrix}.$$

Next, strict preference relation P is built, obtained on the base of (3.3) for $N(x) = 1 - x$ and $A_1 = A_2$ equal to the arithmetic mean. As a result P is of the form $P_{ij} = [A_1(\underline{R}_{ij}, 1 - \overline{R}_{ji}), A_2(\overline{R}_{ji}, 1 - \underline{R}_{ji})]$. Since $\mathscr{A}$ (built with the use of the arithmetic mean) is idempotent, by Theorem 3.7 we see that $P = R^*_{\mathscr{F}}$. The non-dominance interval-valued fuzzy set ND_{IV} based on P (calculated for each row of the matrix P) is of the form

$$ND_{IV} = \{(x_1, [0.5, 0.5]), (x_2, [0.52, 0.76]), (x_3, [0.56, 0.7]), (x_4, [0.56, 0.62])\}.$$

After applying $\mathscr{N}_s$, generated by the standard fuzzy negation $N(x) = 1 - x$, to the interval-valued fuzzy sets ND_{IV} we get

$$\mathscr{N}_s(ND_{IV}) = \{(x_1, [0.5, 0.5]), (x_2, [0.24, 0.48]), (x_3, [0.3, 0.44]), (x_4, [0.38, 0.44])\}.$$

Applying, relation $\preceq_\pi$ to the intervals of $\mathscr{N}_s(ND_{IV})$ we get: $x_i \preceq_\pi x_1$ for $i \in \{1, 2, 3, 4\}$, $x_2 \preceq_\pi x_i$ for $i \in \{1, 2, 3, 4\}$, and similarly $x_3 \preceq_\pi x_i$ for $i \in \{1, 2, 3, 4\}$, $x_4 \preceq_\pi x_i$ for $i \in \{1, 2, 3, 4\}$. Since $\preceq_\pi$ is complete, all alternatives are comparable but not in the unique way. However, we see that the alternative x_1 is the winning one. To make decision about the remaining alternatives, we choose additional method of comparing intervals, i.e. calculating the width of intervals in $\mathscr{N}_s(ND_{IV})$. As a result, taking additionally the smallest width (the smallest uncertainty) as a better solution we obtain the following order of alternatives: $x_2 \prec x_3 \prec x_4 \prec x_1$.

Now, to compare the results, we will order the alternatives using five different linear orders. In the first case we use the order $\leq_{Lex1}$. In the second one we use $\leq_{Lex2}$. Next, we use $\leq_{XY}$ and $\leq_{0.4,0.6}$ obtained from $K_{0.4}(x, y) = 0.4x + 0.6y$ and $K_{0.6}(x, y) = 0.6x + 0.4y$ (cf. Example 1.5). Finally, in the last one, we use $\leq_{0.2,0.7}$ obtained from $K_{0.2}(x, y) = 0.2x + 0.8y$ and $K_{0.7}(x, y) = 0.7x + 0.3y$. The obtained ordering of alternatives, with the given different admissible orders between intervals, are presented in Table 3.1.

With the three orders $\leq_{Lex1}$, $\leq_{XY}$ and $\leq_{0.4,0.6}$ we obtained the same order of alternatives and this the same order as the one obtained for comparing intervals with the relation $\preceq_\pi$ (and additionally taking into account the width of intervals).

The time complexity of the given algorithm is the polynomial one. Interval-valued fuzzy relations $R_1, \ldots, R_n$ are defined on a set X consisting of m elements, so the

Table 3.1 Ordering of alternatives for different admissible orders

Type of order	Obtained order of alternatives
$\leq_{Lex1}$	$x_2 \leq_{Lex1} x_3 \leq_{Lex1} x_4 \leq_{Lex1} x_1$
$\leq_{Lex2}$	$x_3 \leq_{Lex2} x_4 \leq_{Lex2} x_2 \leq_{Lex2} x_1$
$\leq_{XY}$	$x_2 \leq_{XY} x_3 \leq_{XY} x_4 \leq_{XY} x_1$
$\leq_{0.4,0.6}$	$x_2 \leq_{0.4,0.6} x_3 \leq_{0.4,0.6} x_4 \leq_{0.4,0.6} x_1$
$\leq_{0.2,0.7}$	$x_3 \leq_{0.2,0.7} x_4 \leq_{0.2,0.7} x_2 \leq_{0.2,0.7} x_1$

time complexity of the algorithm will depend on the variable m (the size of a matrix representing $R_1, \ldots, R_n$). Determination of the transitivity type of each interval-valued fuzzy relation takes $O(m^3)$ time complexity. Taking into account that the remaining steps in the Algorithm 1 will not exceed this order of complexity we see that the Algorithm 1 takes $O(m^3)$ time complexity.

In this chapter we discussed preservation of pos-B and nec-B-transitivity properties by diverse types of aggregation functions for interval-valued fuzzy settings. For decision making problems we proposed to apply new approaches - depending on the interpretation *possible* or *necessary*. Namely, if we consider the *possible* approach, then relation $\preceq_\pi$ is involved and we proposed to apply in such case the related notions of transitivity and aggregation operators. Analogously, for the *necessary* approach, with the relation $\preceq_\nu$ involved, we proposed to consider the related notions of transitivity and aggregation operators. This approach enables to involve the interpretation of uncertainty intervals in the applied notions.

References

1. Kacprzyk, J.: Group decision making with a fuzzy linguistic majority. Fuzzy Sets Syst. **18**(2), 105–118 (1986)
2. Szmidt, E., Kacprzyk, J.: Using intuitionistic fuzzy sets in group decision making. Control. Cybern. **78**, 183–195 (1996)
3. Szmidt, E., Kacprzyk, J.: A new approach to ranking alternatives expressed via intuitionistic fuzzy sets. In: Ruan, D., et al. (eds.) Computational Intelligence in Decision and Control, pp. 265–270. World Scientific, Singapore (2008)
4. Bentkowska, U.: New types of aggregation functions for interval-valued fuzzy setting and preservation of pos-B and nec-B-transitivity in decision making problems. Inf. Sci. **424**, 385–399 (2018)
5. Bentkowska, U., Pękala, B., Bustince, H., Fernandez, J., Jurio, A., Balicki, J.: N-reciprocity property for interval-valued fuzzy relations with an application to group decision making problems in social networks. Int. J. Uncertain. Fuzziness Knowl. Based Syst. **25**(Suppl. 1), 43–72 (2017)
6. Xu, Z.: On compatibility of interval fuzzy preference relations. Fuzzy Optim. Decis. Mak. **3**, 217–225 (2004)
7. Fodor, J., Roubens, M.: Fuzzy Preference Modelling and Multicriteria Decision Support. Kluwer Academic Publishers, Dordrecht (1994)
8. Montero, J., Tejada, J., Cutello, C.: A general model for deriving preference structures from data. Eur. J. Oper. Res. **98**, 98–110 (1997)
9. Bilgiç, T.: Interval-valued preference structures. Eur. J. Oper. Res. **105**, 162–183 (1998)
10. Barrenechea, E., Fernandez, J., Pagola, M., Chiclana, F., Bustince, H.: Construction of interval-valued fuzzy preference relations from ignorance functions and fuzzy preference relations: application to decision making. Knowl. Based Syst. **58**, 33–44 (2014)
11. Pękala, B.: Uncertainty Data in Interval-Valued Fuzzy Set Theory. Properties, Algorithms and Applications. Studies in fuzziness and soft computing. Springer, Cham (2019)
12. Bentkowska, U., Bustince, H., Jurio, A., Pagola, M., Pękala, B.: Decision making with an interval-valued fuzzy preference relation and admissible orders. Appl. Soft Comput. **35**, 792–801 (2015)
13. Orlovsky, S.A.: Decision-making with a fuzzy preference relation. Fuzzy Sets Syst. **1**(3), 155–167 (1978)

14. Hüllermeier, E., Brinker, K.: Learning valued preference structures for solving classification problems. Fuzzy Sets Syst. **159**, 2337–2352 (2008)
15. Hüllermeier, E., Vanderlooy, S.: Combining predictions in pairwise classification: an optimal adaptive voting strategy and its relation to weighted voting. Pattern Recognit. **43**(1), 128–142 (2010)
16. Bustince, H., Galar, M., Bedregal, B., Kolesárová, A., Mesiar, R.: A new approach to interval-valued Choquet integrals and the problem of ordering in interval-valued fuzzy sets applications. IEEE Trans. Fuzzy Syst. **21**(6), 1150–1162 (2013)
17. Dubois, D.: The role of fuzzy sets in decision sciences: old techniques and new directions. Fuzzy Sets Syst. **184**, 3–28 (2011)

Chapter 4
Optimization Problem of *k-NN* Classifier for Missing Values Case

> *You have to teach your algorithm what it can do and what it cannot do because, otherwise, there is a risk that the algorithms will learn the tricks of the old cartels.*
>
> Margrethe Vestager

In this chapter we present a method of dealing with missing values in data sets. This method uses interval-valued fuzzy calculus and we show that it outperforms other methods that were previously known. The obtained results may be useful in diverse computer support systems but especially in the computer support systems devoted to support the medical diagnosis. In data sets of such systems, for many reasons (e.g. financial, the lack of a specific medical equipment in a given medical center, the high risk of deterioration in a patients health after a potential examination), missing values may appear very often. Moreover, the presented results may be useful not only for the researchers working in the area of interval-valued fuzzy sets but generally for those involved in the methods of interval modeling.

Classifiers also known in literature as *decision algorithms*, *classifying algorithms* or *learning algorithms* may be treated as constructive, approximate descriptions of concepts (decision classes). These algorithms constitute the kernel of *decision systems* that are widely applied in solving many problems occurring in such domains as *pattern recognition*, *machine learning*, *expert systems*, *data mining* and *knowledge discovery* (cf. [1]).

Data sets used for classifiers generation may be represented using data tables. In this representation individual objects correspond to rows of a given data table and attributes to columns of a given data table. In this chapter, we consider *decision tables* of the form $\mathbf{T} = (U, A, d)$ in Pawlak's sense (cf. [2]) for representation of data tables, where A is a set of attributes or columns in the data table, U is a set

U. Bentkowska, *Interval-Valued Methods in Classifications and Decisions*, Studies in Fuzziness and Soft Computing 378,
https://doi.org/10.1007/978-3-030-12927-9_4

of objects (rows) in the data table, and d is a distinguished attribute from the set A called a *decision attribute*.

When classifiers are used, there is a problem of lowering its performance on test data due to the missing values in data test tables. The more missing values the lower quality of the classification can be significant compared to the situation when we classify test objects without missing values. Classifiers are usually created on training data that does not have many missing values or even does not have them at all. As a result, they are able to learn well to recognize test cases based on the values of conditional attributes. Therefore, when classifying test objects with missing values, difficulties arise due to the fact that the classifier cannot recognize the test object.

In literature there can be found descriptions of numerous approaches to constructing classifiers, which are based on such paradigms of machine learning theory as *classical and modern statistical methods*, *neural networks*, *decision trees*, *decision rules*, and *inductive logic programming* (cf. [1, 3–6]).

We propose a new approach of creating uncertainty intervals for the k nearest neighborhood classifier (*k-NN*) and then aggregating the obtained intervals with the use of interval-valued aggregation operators. Our aim is to compare the new interval methods with the classical and generally recognized implementation of *k-NN* classifiers. We compare the results with two versions of classically applied *k-NN* classifiers. The results show that the newly proposed classifier, based on interval modeling of missing values and interval-valued aggregation methods applied, performs much better in situation of increase of missing data than the classical versions of *k-NN* classifiers do.

As for the *k-NN* method, there are many variants and several parameters that can be used (cf. [7, 8]). One of the most important parameters is the distance used, which can be chosen in many ways. We have chosen the Euclidean distance since it is the best known one and the most frequently used. In addition, it matches data with numeric attributes that were analyzed. Moreover, there are many implementations of the *k-NN* method in various software libraries and programming languages. We applied Java programming language, where one of the most important implementations of the *k-NN* algorithm is the implementation in the WEKA API library (cf. [9, 10]) used in the mentioned experiments.

In our new version of the classifier we propose to consider interval modeling in the case of incomplete data. In this way we obtain the so called 'uncertainty intervals' (cf. [11]) used in interval-valued fuzzy settings (cf. [12]). Moreover, aggregation of uncertainty intervals is proposed. Here we present the results on aggregation methods connected with possible and necessary aggregation functions [13]. These notions are connected with possible and necessary comparability relations which follow from the epistemic settings of interval-valued calculus [11, 14].

To assess the quality of the analyzed methods, we used the AUC parameter, which is the area under the ROC curve (cf. [15, 16]). This parameter is much more reliable than giving specific values of sensitivity and specificity, because each such pair of parameters corresponds to one point on the ROC curve, it means that the AUC method gives much more information and enables the adjustment of sensitivity and specificity. Certainly, the use of this method requires classifiers with the ability to adjust sensitivity and specificity, but classifiers that were used in our experiments have such an option.

An evaluation of our approach was conducted on data sets from a machine learning repository UCI [17]. The results confirmed that methods based on interval modeling and interval-valued aggregation make it possible to reduce the negative impact of lack of data.

In the experiments we considered methods of improving the quality of classification by the k nearest neighborhood classifiers in the case of large number of missing values. It was applied the classical version of the classifier, the classifier with the aggregation of certainty coefficients of individual classifiers with the use of the arithmetic mean, as well as new versions of the classifier with the use of diverse interval-valued aggregation operators. After adding the missing values to the data, the quality of the classical method is going down. However, the interval modeling of missing values and application of interval-valued aggregation operators in the classifiers cause a much slower decrease of their quality. Details of the performed experiments will be presented in the next sections.

4.1 Construction of the Classifiers

We would like to pay special attention to the usefulness of interval modeling while dealing with the missing values in classification process and application of interval-valued aggregation operators in improving the classification results. Especially, we consider aggregation operators from the classes of possible and necessary aggregation functions.

4.1.1 Aggregation Operators for Interval-Valued Settings

Below there are given six aggregation operators applied in our experiments. Some of the formulas were given in Chap. 2 but to make it transparent we present here explicitly the whole group of applied in the current experiment aggregation operators. These are the following aggregation operators.

$$\mathscr{A}_1(\mathbf{x}_1, \mathbf{x}_2, \ldots, \mathbf{x}_n) = \left[\frac{\underline{x}_1 + \underline{x}_2 + \cdots + \underline{x}_n}{n}, \frac{\overline{x}_1 + \overline{x}_2 + \cdots + \overline{x}_n}{n}\right], \tag{4.1}$$

where $\mathscr{A}_1$ is a representable interval-valued aggregation function constructed with the use of the arithmetic mean (this is the interval-valued arithmetic mean, cf. Example 2.13). It is also a possible and a necessary aggregation function (cf. Corollary 2.2 and Theorem 2.9).

$$\mathscr{A}_2(\mathbf{x}_1, \mathbf{x}_2, \ldots, \mathbf{x}_n) = \Bigg[\frac{\underline{x}_1 + \underline{x}_2 + \cdots + \underline{x}_n}{n},$$
$$\max\left(\frac{\underline{x}_1 + \overline{x}_2 + \cdots + \overline{x}_n}{n}, \frac{\overline{x}_1 + \underline{x}_2 + \overline{x}_3 + \cdots + \overline{x}_n}{n}, \ldots, \frac{\overline{x}_1 + \cdots + \overline{x}_{n-1} + \underline{x}_n}{n}\right)\Bigg], \tag{4.2}$$

where $\mathscr{A}_2$ is an aggregation operator of the form (2.37) and it is also a necessary aggregation function (cf. Proposition 2.6) but it is not a pos-aggregation function (cf. Example 2.35).

$$\mathscr{A}_3(\mathbf{x}_1, \mathbf{x}_2, \ldots, \mathbf{x}_n) = \left[\frac{\underline{x}_1 + \underline{x}_2 + \cdots + \underline{x}_n}{n}, \frac{\overline{x}_1^2 + \overline{x}_2^2 + \cdots + \overline{x}_n^2}{\overline{x}_1 + \overline{x}_2 + \cdots + \overline{x}_n}\right] \tag{4.3}$$

and its more general version $\mathscr{A}_4$, where $p \in \mathbb{N}$, $p \geqslant 2$,

$$\mathscr{A}_4(\mathbf{x}_1, \mathbf{x}_2, \ldots, \mathbf{x}_n) = \left[\frac{\underline{x}_1 + \underline{x}_2 + \cdots + \underline{x}_n}{n}, \frac{\overline{x}_1^p + \overline{x}_2^p + \cdots + \overline{x}_n^p}{\overline{x}_1^{p-1} + \overline{x}_2^{p-1} + \cdots + \overline{x}_n^{p-1}}\right], \tag{4.4}$$

where $\mathscr{A}_3$ and $\mathscr{A}_4$ are possible aggregation functions (cf. Example 2.16) but they are neither interval-valued aggregation functions (cf. Theorem 2.7) nor nec-aggregation functions (cf. Theorem 2.9). In the case of $\mathscr{A}_4$ the greater the p value, the greater the value of the end of interval (which depends on the value p, cf. (2.17)). $\mathscr{A}_3, \mathscr{A}_4$ are given with the convention $\frac{0}{0} = 0$, where $\frac{0}{0}$ occurs in the case $\mathbf{x}_1 = \mathbf{x}_2 = \cdots = \mathbf{x}_n = [0, 0]$.

$$\mathscr{A}_5(\mathbf{x}_1, \mathbf{x}_2, \ldots, \mathbf{x}_n) = \left[\sqrt{\frac{\underline{x}_1^2 + \underline{x}_2^2 + \cdots + \underline{x}_n^2}{n}}, \sqrt[3]{\frac{\overline{x}_1^3 + \overline{x}_2^3 + \cdots + \overline{x}_n^3}{n}}\right], \tag{4.5}$$

$$\mathscr{A}_6(\mathbf{x}_1, \mathbf{x}_2, \ldots, \mathbf{x}_n) = \left[\sqrt[3]{\frac{\underline{x}_1^3 + \underline{x}_2^3 + \cdots + \underline{x}_n^3}{n}}, \sqrt[4]{\frac{\overline{x}_1^4 + \overline{x}_2^4 + \cdots + \overline{x}_n^4}{n}}\right], \tag{4.6}$$

where $\mathscr{A}_5$ and $\mathscr{A}_6$ are interval-valued representable aggregation functions (constructed with the use of adequate root-power means) and they are also possible aggregation functions (cf. Corollary 2.2) but they are not necessary aggregation functions (cf. Theorem 2.9). We used some promising examples of aggregation operators to be applied in classification problems. Since this is the first attempt to apply the new classes of aggregation operators in classification problems, we have chosen some typical representatives of both the classically used aggregation operators for interval-valued fuzzy settings (cf. Definition 2.15) and the new classes of aggregation operators such as possible and necessary aggregation functions (cf. Definitions 2.20 and 2.21). The aggregation operators that were chosen are intuitively suitable for the construction of classifiers for the considered here type of data.

4.1.2 Missing Values in Classification

Now, we present the motivation to consider intervals when missing values appear in classification procedure. This also justifies the application of aggregation operators for interval calculus. The problem of missing values in classification problems is usually solved in the following three ways.

The first method is that the missing value is treated as a normal value that carries certain information. For example, for medical data, the missing value for a certain attribute representing the result of a specialized medical examination often means that this test was not needed because of the patient's condition. It is therefore information about the patient's condition. However, this approach requires teaching the classifier on data containing this kind of missing values. The classifier so learned can classify test objects containing missing values as well. This approach does not really consider missing values, because they are treated as normal values.

The second approach is that classifiers that can classify objects only on the basis of those attributes on which there are no empty spaces are created. This approach requires the creation of specific classifiers that can classify test objects based only on some attributes. A typical example of such a classifier is a classifier based on decision rules. If there are many rules in a set of rules, then usually one of them is able to recognize the test object based on non-missing values. It is worth noting, that such a classifier is working much worse if it classifies test objects with missing values compared to the case when there are no such values in the test object.

The third approach, which seems to be the most commonly used in practice (cf. [18, 19]), is that when classifying a test object having empty spaces, before classifying the object, empty spaces are filled with a specific value determined based on training data. It is only after this filling that the classifier is used to classify the test object thus completed. The method of value replenishment (data imputation) usually consists in searching for the most frequently occurring value of a symbolic attribute or the average value of a numerical attribute. Although this approach seems uncertain, in practice it often allows good classification results to be obtained on test data with empty spaces. It is worth noting, however, that this method in a sense introduces artificial values into data, which in reality could not occur at this point. These are frequent values in the training data and therefore the method works well for typical objects. However, for untypical test objects, the classifier used in this way may be mistaken. Another method was applied in OvaExpert, one of the real-life diagnosis support systems for ovarian tumor, where the missing values were replaced with the whole intervals representing the given features (cf. [20–28]). This approach proved to be effective to make high-quality decisions under incomplete information and uncertainty.

If we assume that the phenomenon of appearance of missing values in the test data is random, it means that the biggest problems with the correct classification of test data may have classifiers by definition using all conditional attributes. This is due to the fact that classifiers that do not use all attributes for classification at the same time (e.g. decision rules, decision trees, etc.) when classifying a test object can often use attribute values that do not require the above-mentioned fill and therefore classify objects more confidently. However, classifiers that use all attributes do not have this chance and must also use artificially filled values. A typical example of such a classifier is the k-nearest neighbors classifier (*k-NN*) [1]. The test object can be classified using the *k-NN* method with diverse parameters k. Therefore, a conflict appears between classifiers that operate on the basis of different k values, which must be resolved in order to finally classify the test object. In this work, to resolve this conflict, we suggest aggregation of uncertainty intervals by individual classifiers. As a result, we build a new compound classifier. There are applied classical versions of the classifier as well as the aggregated versions with diverse interval-valued aggregation operators. We compare the obtained results for the classifier using interval modeling of missing data with the classical version of the classifier and the classifier with the aggregation of certainty coefficients of individual classifiers by means of the arithmetic mean (without the use of uncertainty intervals). We will show that appearance of the missing values causes the decrease of the quality of the classical methods. However, adding missing values to the aggregated versions of classifiers (with interval modeling of missing values) causes a much slower decrease of their quality and outperform the other two presented methods which is confirmed by statistical measures.

4.1.3 k-NN Algorithm

In an unpublished US Air Force School of Aviation Medicine report, Fix and Hodges in 1951 introduced a non-parametric method for pattern classification that has since become known the k-nearest neighbor rule [29]. In 1967, some of the formal properties of the k-nearest neighbor rule were obtained [30]. The *k-NN* (cf. [1]) is a method for classifying objects based on the closest training examples in a feature space. It is a type of instance-based learning. An object is classified by a majority vote of its neighbors, with the object being assigned to the class most common among its k nearest neighbors (k is a positive integer, usually small). If $k = 1$, then the object is simply assigned to the class of its nearest neighbor. There were considered also distance weighted approaches ([7, 8]), soft computing methods [31] and fuzzy ones ([32, 33]).

We are using three methods involving k-nearest neighbors classifier. The first, is the classical one and it is denoted as Algorithm C, for short C. The experiments were performed with the use of WEKA API library (cf. [9, 10]), where before calculating the Euclidean distance, for the purpose of *k-NN* algorithm (Algorithm C) values on all numeric attributes are normalized to the interval [0, 1]. The difference between numeric values in the distance formula is calculated as the absolute value between them and for nominal values the result equals 0 if they are the same and it equals 1 if

they are different. The case of missing values is solved at the stage of calculating the
distance between the values on a given attribute. For nominal attributes, assume that
a missing feature is maximally different from any other feature value. Thus if either
or both values are missing, or if the values are different, the difference between them
is taken as one; the difference is zero only if they are not missing and both are the
same. For numeric attributes, the difference between two missing values is also taken
as one. However, if just one value is missing, the difference is often taken as either the
(normalized) size of the other value or one minus that size, whichever is larger. This
means that if values are missing, the difference is as large as it can possibly be. The
described algorithm of determining the distance between test objects is presented in
Algorithm 2.

Algorithm 2: Distance between two objects in *k-NN* (WEKA API library)

Input:
1. training data set represented by decision table $\mathbf{T} = (U, A, d)$, where $n = card(U)$ and $l = card(A)$,
2. given k
3. tested objects u and v.

Output: Distance between objects u and v

1 **begin**
2   **if** *The attributes u_j and v_j for $j \in \{1, \ldots, l\}$ are nominal* **then**
3   **if** *The value of attribute u_j and the value v_j are both non-missing and the same* **then**
4 | The difference between u_j and v_j is equal to 0.
5   **else**
6 | The difference between u_j and v_j is equal to 1.
7   **end**
8   **else**
9   **if** *The attributes u_j and v_j for $j \in \{1, \ldots, l\}$ are numeric* **then**
10   **if** *The value of attribute u_j and the value v_j are both non-missing* **then**
11 | The difference between u_j and v_j is equal to $|u_j - v_j|$.
12   **else**
13   **if** *Both u_j and v_j are missing* **then**
14 | The difference between them is equal to 1.
15   **else**
16 | The difference between u_j and v_j is equal to $\max(v_j, 1 - v_j)$ assuming that v_j is the non-missing one.
17   **end**
18   **end**
19   **end**
20   **end**
21 **end**

The test object data can be classified using the *k-NN* method with diverse parameters k. For example, it can be $k = 1, 3, 5, 10, 20$ etc. In each of these cases a different set of neighbors is used, which can classify the test object with different levels of confidence for different decision classes. This is due, among other things, to the fact that various objects appear in particular environments, including objects with artificially filled values. Therefore, a conflict appears between classifiers that operate on the basis of different k values, which must be resolved in order to finally classify the test object. In this work, to resolve this conflict, we suggest aggregation of uncertainty coefficients by individual classifiers. This is connected with the second method, Algorithm M (for short M, cf. Algorithm 3) which is based on the fact that we have a number of classical *k-NN* classifiers and we make a simple aggregation (with the use of the arithmetic mean) of the degree of classification from all the individual classifiers.

Algorithm 3: Classification of a tested object by the M classifier

---

Input:
1. training data set represented by decision table $\mathbf{T} = (U, A, d)$, where $n = card(U)$ and $l = card(A)$,
2. collection $C_1, ..., C_m$ of *k-NN* classifiers for different k, where, e.g., $k \in \{5, 10, 20, 30\}$,
3. tested object u.

Output: The membership of the object u to the "main class" or to the "subordinate class"

begin
  for $i := 1$ **to** m **do**
    Compute certainty coefficient ("main class" membership probability) for the given tested object u using the classifier C_i and assign it to p_i
  end
  Determine the final certainty coefficient p for the object u by aggregating (with a use of the arithmetic mean) the certainty coefficients $p_1,...,p_m$.
  if $p > 0.5$ **then**
    return *u belongs to the "main class"*
  else
    return *u belongs to the "subordinate class"*
  end
end

We suggest also another method - Algorithm F involving interval-valued aggregation of uncertainty intervals by individual classifiers. Details of this method are given in the next section.

4.1.4 New Version of Classifier

The proposed new method of building a classifier, Algorithm F (for short F, cf. Algorithm 4) consists in using a special procedure that is in fact a complex classifier.

At the beginning, we set a certain decision class, which we call the main class. We assume that there is a second decisional class in the data set, which we call a subordinate class. The main class can be, for example, the class of patients suffering from a given disease, and the subordinate class may be the class of healthy patients. For a given family of k parameters (e.g. $k = 1, 3, 5, 10$), each *k-NN* classifier determines the uncertainty interval of belonging of a test object to the main class (it will be explained later). Then, the method of aggregating intervals is used to obtain the final uncertainty interval. The final uncertainty interval is used to classify the object at the level of uncertainty corresponding to the center point of the final uncertainty interval.

Now, we explain the method of determining the uncertainty interval by a single *k-NN* classifier. The method proposed for determining the uncertainty interval by a single classifier consists in the fact that a test object having missing values is classified by a given *k-NN* classifier in a specific way. Namely, during this classification, many classifications of different objects are actually made, which are constructed based on the test object. The construction of these objects is based on the fact that missing values in the object are filled in various ways based on the values from the training data. The ideal situation here would be that during the classification procedure, all possible test objects that can be generated from a given test object are classified by inserting empty values in all possible ways of the attribute values from the training data for the given attribute. The result of each such classification is the value of the certainty of belonging to the main class. Thanks to this value, the confidence interval may be computed by determining the minimum of these values (lower end of the interval) and the maximum (upper end of the interval) (cf. [21, 34]). Unfortunately, due to the possibility of the appearance of a very large number of objects generated from a given test object, in practice, the above method cannot be used in its pure form. If, for example, there are 20 attributes in the training table and each of them can have only 10 values, then in the case of the classification of a test object having 10 empty and 10 filled places, $10^{10} = 10$ billion objects would have to be classified. From the point of view of computational complexity, this is definitely too much. Therefore, we propose the Monte Carlo method of choosing the above objects. This method consists in the fact that in the space of all possible objects that can be generated for a given test object with missing values, we select a random sample (the draw being made in accordance with the distribution of variable). Then, we classify only objects from this sample and on the basis of the obtained classification results, we estimate the lower and upper end of the classification uncertainty interval.

Now, we present a brief analysis of the computational time complexity of the Algorithm 4. If we assume that the classical *k-NN* algorithm works in time of order $O(n \cdot l)$, where n is the number of objects in the set U, and l is the number of attributes in the set A, then it is easy to see that the time complexity of the Algorithm 4 is of order

Algorithm 4: Classification of a tested object by the F classifier

Input:
1. data set represented by decision table $\mathbf{T} = (U, A, d)$, where $n = card(U)$ and $l = card(A)$,
2. collection $C_1, ..., C_m$ of *k-NN* classifiers for different k, where, e.g., $k \in \{5, 10, 20, 30\}$,
3. fixed parameter r, e.g., $r = 10$,
4. aggregation function $\mathscr{A}$,
5. tested object u

Output: The membership of the object u to the "main class" or to the "subordinate class"

begin
  if *exists at least one missing value in the object* u **then**
    for $i := 1$ **to** m **do**
      Choose randomly with the Monte Carlo method r objects $u_1, ..., u_r$ on the basis of object u, where any object u_j ($j \in \{1, ..., r\}$) is constructed in the following way:
      begin
        Copy values of attributes from u to u_j;
        For each attribute whose value in u_j is missing, replace it with a randomly selected value from the range of possible values for this attribute (the range designated from the training data);
      end
      Compute certainty coefficient for objects $u_1, ..., u_r$ using the classifier C_i and assign these values to $p_1, ..., p_r$;
      Compute $min\{p_1, ..., p_r\}$ and assign it to min_i;
      Compute $max\{p_1, ..., p_r\}$ and assign it to max_i;
    end
    Determine the uncertainty interval $[down(u), up(u)]$ for the object u by aggregating (with the use of interval-valued aggregation function $\mathscr{A}$) the intervals $[min_1, max_1]$, ..., $[min_m, max_m]$;
    Determine the final certainty coefficient $\mathbf{p} = \frac{down(u)+up(u)}{2}$ for the object u;
  else
    for $i := 1$ **to** m **do**
      Compute certainty coefficient ("main class" membership probability) for the given tested object u using the classifier C_i and assign it to p_i ;
    end
    Determine the uncertainty interval $[down(u), up(u)]$ for the object u by aggregating (with the use of $\mathscr{A}$) the intervals $[p_1, p_1]$, ..., $[p_m, p_m]$;
    Determine the final certainty coefficient $\mathbf{p} = \frac{down(u)+up(u)}{2}$ for the object u;
  end
  if $\mathbf{p} > 0.5$ **then**
    return u *belongs to the "main class"*;
  else
    return u *belongs to the "subordinate class"*;
  end
end

Table 4.1 Experimental data set details

UCI data	Objects	Attributes	Classes
Banknote	1,372	5	2
Biodeg	1,055	41	2
Breast cancer	699	10	2
Diabetes	768	9	2
German	1,000	25	2
Ozone	2,536	73	2
Parkinson	1,040	29	2
Red wine	1,599	12	2
Rethinopathy	1,151	20	2
Spam	4,601	57	2

$O(m \cdot r \cdot n \cdot l)$, where m is the number of classifiers from the collection $C_1, \ldots, C_m$ and r is the parameter of the Monte Carlo method. This complexity is not large and the algorithm can be used in practical applications wherever the *k-NN* algorithm can be used.

4.2 Experiments

Now, we present detailed information about the experiments that were carried out (cf. [35]). The data sets are described along with the training and test phase details as well as the concept of AUC, namely the area under the ROC curve, is recalled.

4.2.1 Conditions of Experiments

The experiments have been performed on the 10 data sets obtained from UC Irvine (UCI) Machine Learning repository [17]. Table 4.1, shows the summary of the characteristics of the data sets. The considered attributes have numerical values only.

Here, we provide a brief description of the considered data sets.

- Banknote - data were extracted from images that were taken for the evaluation of an authentication procedure for banknotes. Data were extracted from images that were taken from genuine and forged banknote-like specimens. For digitization, an industrial camera usually used for print inspection was used. The final images have 400 × 400 pixels. Due to the object lens and distance to the investigated object gray-scale pictures with a resolution of about 660 dpi were gained. Wavelet Transform tool were used to extract features from images. There were five attributes: variance of wavelet transformed image (continuous), skewness of wavelet transformed image (continuous), curtosis of wavelet transformed image (continuous), entropy of image (continuous), class (integer).

- Biodeg (QSAR Biodegradation Data Set, cf. [36]) - data set containing values for 41 attributes (molecular descriptors) used to classify 1,055 chemicals into 2 classes (ready and not ready biodegradable). The data have been used to develop QSAR (Quantitative Structure Activity Relationships) models for the study of the relationships between chemical structure and biodegradation of molecules.
- Breast cancer (Breast Wisconsin Original Data Set) - the creator is Dr. William H. Wolberg (physician, University of Wisconsin Hospitals Madison, Wisconsin, USA, [37]). The classification is into classes *benign* or *malignant*.
- Diabetes - Diabetes patient records were obtained from two sources: an automatic electronic recording device and paper records.
- German (German Credit Data) - classifies people described by a set of attributes as good or bad credit risks.
- Ozone (Ozone Level Detection Data Set) - two ground ozone level data sets are included in this collection. One is the eight hour peak set, the other is the one hour peak set. Those data were collected from 1998 to 2004 at the Houston, Galveston and Brazoria area. The following are specifications for several most important attributes that are highly valued by Texas Commission on Environmental Quality (TCEQ): O 3 - local ozone peak prediction, Upwind - upwind ozone background level, EmFactor - precursor emissions related factor, Tmax - maximum temperature in degrees Fahrenheit, Tb - base temperature where net ozone production begins (50 Fahrenheit), SRd - solar radiation total for the day, WSa - wind speed near sunrise (using 09-12 UTC forecast mode), WSp - wind speed mid-day (using 15-21 UTC forecast mode). More details can be found in the relevant paper [38].
- Parkinson (Parkinson Speech Dataset with Multiple Types of Sound Recordings Data Set) - the training data belongs to 20 Parkinson's Disease (PD) patients and 20 healthy ones. From all patients, multiple types of sound recordings are taken.
- Red wine (Wine Quality Data Set) - two data sets are included, related to red and white *vinho verde* wine samples, from the north of Portugal; the goal is to model wine quality based on physicochemical test.
- Retinopathy (Diabetic Retinopathy Debrecen Data Set) - this dataset contains features extracted from the Messidor image set to predict whether an image contains signs of diabetic retinopathy or not. This dataset contains features extracted from the Messidor image set to predict whether an image contains signs of diabetic retinopathy or not. All features represent either a detected lesion, a descriptive feature of a anatomical part or an image-level descriptor. The binary result of quality assessment is bad quality or sufficient quality. The underlying method image analysis and feature extraction as well as the classification technique is described in [39].
- Spam (Spambase Data Set) - the collection of spam e-mails coming from postmaster and individuals who had filed spam. The collection of non-spam e-mails coming from filed work and personal e-mails, and hence the word 'george' and the area code '650' are indicators of non-spam. These are useful when constructing a personalized spam filter. One would either have to blind such non-spam indicators

or get a very wide collection of non-spam to generate a general purpose spam filter. The last column of Spam denotes whether the e-mail was considered spam (1) or not (0), i.e. unsolicited commercial e-mail. Most of the attributes indicate whether a particular word or character was frequently occuring in the e-mail. The run-length attributes (55–57) measure the length of sequences of consecutive capital letters. Creators are Mark Hopkins, Erik Reeber, George Forman, Jaap Suermondt from Hewlett-Packard Labs, 1501 Page Mill Rd., Palo Alto, CA 94304.

To measure the quality of a binary classifier, the first step is to identify the types of errors that it can make. The classifier indicates, on the basis of the input data, that the object belongs to two classes: positive (P) and negative (N) (cf. [40]). To assess the overall quality of the analyzed methods the AUC parameter was used. The use of AUC method requires classifiers with the ability to adjust sensitivity (SN) called also true positive rate (TPR) and specificity (SP) called also true negative rate (TNR), where

$$SN = TPR = \frac{TP}{P} = \frac{TP}{TP + FN},$$

$$SP = TNR = \frac{TN}{N} = \frac{TN}{TN + FP},$$

$$FPR = \frac{FP}{N} = \frac{FP}{FP + TN},$$

$$FNR = \frac{FN}{P} = \frac{FN}{FN + TP}.$$

TP - True Positive specifies the number of positive class objects that have been classified as positive. TN - True Negative specifies the number of negative class objects that have been classified as negative. FP - False Positive specifies the number of negative class objects that have been classified as positive. FN - False Negative specifies the number of positive class objects that have been classified as negative. Sensitivity (SN) specifies how many of the positive class objects are properly categorized. It may be interpreted as the probability that the classification will be correct, provided that the case is positive. Specificity (SP) shows how often the model correctly classifies objects from the negative class. In other words, it is the probability that the classification will be correct, provided that the case was negative. FPR is called type I error or α error and FNR is called type II error or β error. Important dependencies between the given measures are presented below

$$FPR = 1 - SP,$$

$$FNR = 1 - SN.$$

AUC (cf. Fig. 4.1) is the indicator of the quality of a classifier which is the area under the ROC curve (Receiver Operating Characteristic curve, cf. [15, 16]).

Fig. 4.1 ROC and AUC

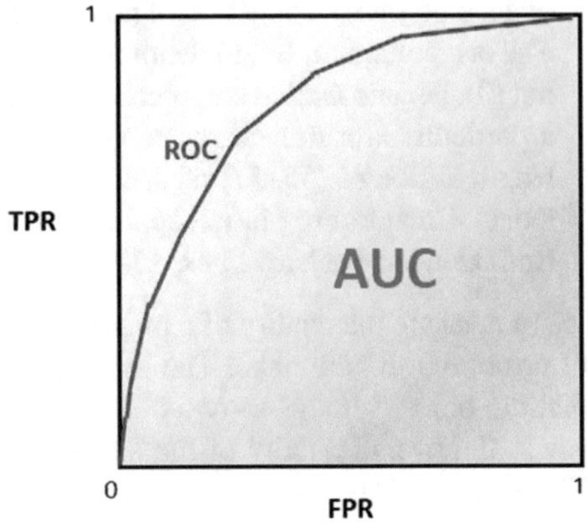

ROC shows the dependence of sensitivity SN on error of the first type FPR during calibration of the classifier (at various threshold settings). These two coefficients are determined on the basis of *cost matrix* (explaining briefly, the matrix presenting a cost of classifying an example of class j as class i) and as a result each singular binary classifier may be presented as a point $(1 - SP, SN)$ in the coordinate system. The bigger is the AUC value the better is the classifier. For the ideal classifier the value AUC is equal to 1.

Details of the performed experiments are presented below.

- All algorithms were implemented and tested in Java programming language (Weka API library, cf. [9, 10]).
- The Euclidean metric is applied for measuring distances.
- Each data set is divided into training and test parts, in the proportion of 50% to 50%.
- Training data does not have missing values.
- A random missing value with different probability is entered into the test data (this is the experiment parameter). In each experiment, missing values are randomly entered into all conditional attributes in the same proportion. For example, if the input parameter is 0.2, then 20% of the missing value is randomly entered into every attribute (column).
- In the experiments, in the method F, six interval-valued aggregation operators $\mathscr{A}_1$–$\mathscr{A}_6$ were applied ($\mathscr{A}_4$ is used with $p = 3$, cf. (4.1)–(4.6)).
- Experiments are carried out for the following classifiers:
  - the classifier with **C** code - classical classifier with *k-NN* method (classifiers for $k = 1, 3, 5, 10, 15, 20$ and 30 were used);
  - the classifier with **M** code - uses the aggregation of certainty coefficients of individual classifiers by means of the arithmetic mean (cf. Algorithm 3);
  - the classifier with **F** code - the algorithm of aggregation of uncertainty intervals described in Sect. 4.1.4 (cf. Algorithm 4).

- Each experiment is repeated 10 times and the average AUC and standard deviation are reported.
- In each experiment, the main class has always been determined, which is given as an experiment parameter. The data was so selected that only two decision classes were considered.

4.2.2 Discussion and Statistical Analysis of the Results

Generally, the increase of missing values resulted in decrease of classification possibility. However, applying the classification method F, with the proposed here method of treating the case of missing values and aggregation of uncertainty intervals, always gave the best classification results. The levels of missing values were: $0, 0.01, 0.03, 0.05, 0.1, 0.2, 0.3, 0.4, 0.5$. In Tables 8.1–8.3 we present the results of experiments for all considered data sets but just with the selected levels of missing values: $0, 0.01, 0.05, 0.1, 0.3, 0.5$.

We now provide an analysis of the tested data sets and performance of the considered here methods. Let us look at the results for the Parkinson data set. Starting from the level 0.05 of missing values the method F was the winning one. If it comes to the performance of aggregation operators it depended on the level of missing values. However, operator $\mathscr{A}_5$ obtained the worst results (although in most of the cases still better than methods C or M). Banknote data set - starting from the level 0.01 of missing values and ending with the level 0.3 the method F was the wining one. Aggregation operators $\mathscr{A}_1$–$\mathscr{A}_4$ or $\mathscr{A}_6$ had always the 5 top results (their position depended on the level of missing values). The aggregation operator $\mathscr{A}_5$ always lose with the method M or C (it seems that $\mathscr{A}_5$ was not appropriate for the aggregation of this set of data). With the level of missing values 0.5 the best was method M but the method F with the aggregation operator $\mathscr{A}_6$ was at the second place. In the case of no missing values, the classification method C was the winner (for $k = 15, 20, 30$), the next was method M, and then the method F with $\mathscr{A}_1$.

Biodeg data set - independently of the level of missing values the method F was the best one (with one of the aggregation operators applied). In the most cases of levels of missing values, five of the aggregation operators $\mathscr{A}_1$–$\mathscr{A}_4$ or $\mathscr{A}_6$ had the top 5 positions. Again the performance of the method F with $\mathscr{A}_5$ was worse comparing to other aggregation operators. Similar situation was with the data sets Red wine, Rethinopathy, Breast cancer and Diabetes, however in the last three data sets sometimes in the sequence of the top methods (for diverse levels of missing values), among the winning methods F, appeared the method C or M. For data sets German, Ozone and Spam, again among the winning methods were F ones with the most frequent appearance but in these cases the performance of $\mathscr{A}_5$ was much better and for the Ozone data set it was one of the best ones.

We see that the obtained results justify our approach of using the method F with the proposed idea of filling missing values together with aggregation of the obtained in this way intervals (with the use of interval-valued aggregation operators). Moreover, we see that the choice of the applied aggregation operator depends on the data set and also on the level of missing values in it. Furthermore, we may notice that it is worth to use diverse types of aggregation operators, not only the representable interval-valued aggregation function $\mathscr{A}_1$ which is a natural extension of the arithmetic mean from [0, 1] to L^I. In most of the cases (regarding data sets and missing value levels) other than $\mathscr{A}_1$ aggregation operators obtained the top results. Finally, high values of AUC confirm good performance of the applied method F in comparison to the methods C and M.

Moreover, we may see that in the method of dealing with missing values used in WEKA API (cf. Sect. 4.1.3), there is obtained the maximum possible range of uncertainty interval. Whereas, thanks to the method applied in Algorithm F we may obtain better results of classification, since the obtained uncertainty intervals are based on real values that may appear in the test objects. In our method the uncertainty intervals are received as a result of drawing values from the training set.

Below we provide the statistical justification of the above given observations. This justification was performed with the use of Statistica program. The obtained results prove that the method F (with diverse aggregation operators involved) significantly outperforms the method C and are much better than the method M. This was verified with the Kruskal-Wallis statistical test (we applied this test since in the data there is no normal distribution of variables and there are more than two compared groups). The results are given in Table 4.2 and for comparing in pair in Table 4.3. We see that for each level of missing values, apart from the level 0.00, the AUC values are significantly higher for the F method comparing to the C method (for the M method this is not the case but the values AUC for the F method are much higher than the M method).

The obtained results depend also on the aggregation operators used in the F method and parameters k used in the C method. Moreover, the values of AUC are decreasing while the levels of missing values are increasing. It was verified (cf. Table 4.4) with the Pearson's correlation coefficient which measures the linear dependence between the variables and the Spearman's rank correlation coefficient which

Table 4.2 Results of the Kruskal-Wallis test

Level of missing values	The value of K-W	p
0	2.770	0.250
0.01	9.990	0.007
0.03	15.400	0.001
0.05	18.120	0.000
0.1	15.600	0.000
0.2	20.800	0.000
0.3	23.800	0.000
0.4	22.800	0.000
0.5	19.900	0.003

Table 4.3 The p value for multiple repetitions in the Kruskal-Wallis test

	Level of missing values	F	M	C
F	0.0		1.000000	0.331160
M	0.0	1.000000		1.000000
C	0.0	0.331160	1.000000	
F	0.01		0.670916	0.005062
M	0.01	0.670916		1.000000
C	0.01	0.005062	1.000000	
F	0.03		0.654580	0.000264
M	0.03	0.654580		1.000000
C	0.03	0.000264	1.000000	
F	0.05		0.564358	0.000063
M	0.05	0.564358		1.000000
C	0.05	0.000063	1.000000	
F	0.1		0.290960	0.000277
M	0.1	0.290960		1.000000
C	0.1	0.000277	1.000000	
F	0.2		0.293146	0.000017
M	0.2	0.293146		1.000000
C	0.2	0.000017	1.000000	
F	0.3		0.153248	0.000004
M	0.3	0.153248		1.000000
C	0.3	0.000004	1.000000	
F	0.4		0.628647	0.000005
M	0.4	0.628647		0.669703
C	0.4	0.000005	0.669703	
F	0.5		1.000000	0.000032
M	0.5	1.000000		0.251136
C	0.5	0.000032	0.251136	

Table 4.4 The measure of correlation between levels of missing values and AUC

Pearson correlation coefficient		Spearman correlation coefficient	
Method	Coefficient	Method	Coefficient
F	−0.6493	F	−0.700769
M	−0.5890	M	−0.673308
C	−0.6000	C	−0.698155

measures any monotonic dependence. In both cases we obtained negative correlation. However, the dependence was stronger in the case of Spearman's rank correlation coefficient.

We have proved that there are linear dependencies between the values AUC and levels of missing values. There are statistically significant changes between the F method and the M method and analogously, between the M method and the C method (we performed the analysis for the F and C methods with adequate parameters). It was applied the test to check if the gradients of the lines are equal (i.e. the drops are parallel). The test has shown that on the level of $p < 0.05$ such equality holds. The gradients of lines are the same but the line representing the F method is significantly higher than the lines representing C and M methods and the line representing the M method is significantly higher than the line representing the C method (it was verified with the test of equality for y-intercepts of the lines which gave negative results). The highest and significantly different values are obtained for the F method (the F method gives on average a higher AUC value than the C method approximately by 0.05). Moreover, the statistically significant changes of values AUC (regarding the levels of missing values) between pairs of methods were confirmed with the Wilcoxon test. In Fig. 4.2 we present linear regressions of changes for average AUC levels depending on the levels of missing values.

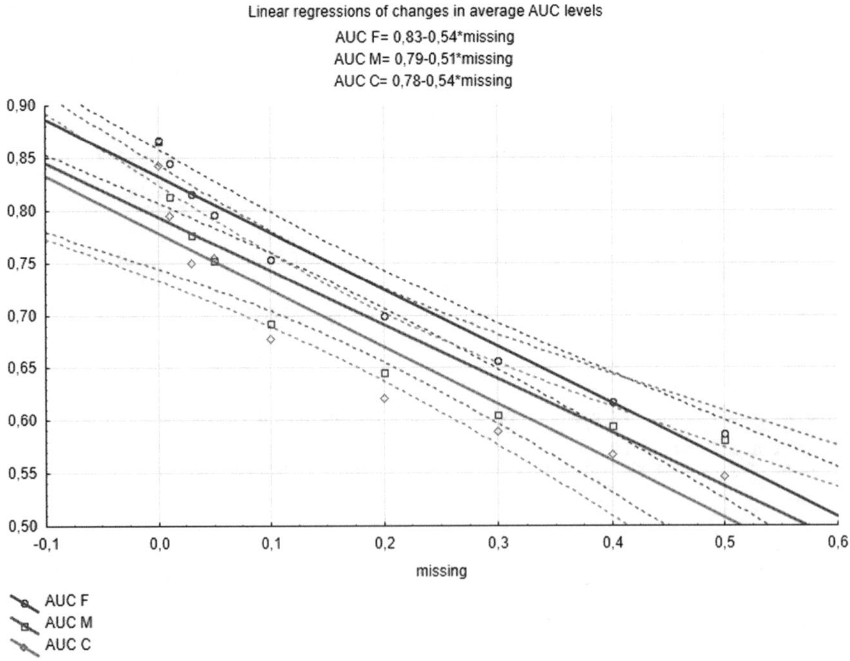

Fig. 4.2 Changes in average AUC depending on the level of missing values

Table 4.5 The best aggregation operators for the F method depending on the level of missing values

Level of missing values	The best aggregation for the F method
0	$\mathscr{A}_4$
0.01	$\mathscr{A}_6$
0.03	$\mathscr{A}_3$ or $\mathscr{A}_6$
0.05	$\mathscr{A}_1$
0.1	$\mathscr{A}_3$
0.2	$\mathscr{A}_2$
0.3	$\mathscr{A}_6$
0.4	$\mathscr{A}_6$
0.5	$\mathscr{A}_2$

We have also checked the behavior of the F method depending on the aggregation operators applied. The Kruskal-Wallis statistical test shows that there are no statistically significant differences between the AUC values for the F method with respect to the aggregation operators applied for each level of missing values. The aggregation operator $\mathscr{A}_5$ obtained worse results than other operators but these are not statistically significant values. In Table 4.5 we provide the best interval-valued aggregation operators (applied in the F method) depending on the levels of missing values.

In Fig. 4.3 we present the method F with the best results regarding the aggregation operators applied in comparison to the method C and the method M. We see that the gradient of the line representing the method F is unchanged and the difference between the F and the C method increased to 0.06.

Moreover, we noticed that the exponential models are better than the linear ones for matching the AUC changes versus levels of missing values. In Fig. 4.4 we present the F method in comparison to the method C and the method M using exponential functions.

In this chapter it was shown that, despite appearance of missing values, application of the interval modeling and aggregation methods cause much slower decrease of the classification quality comparing to the classical versions of *k-NN* classifiers (and classical implementations of *k-NN* such as WEKA API library, cf. Sect. 4.1.3). Application of diverse aggregation operators enabled to choose the one which is the most suitable for a given data set and level of missing values. Moreover, a new method of dealing with missing values was proposed with the satisfactory results. The results of experiments are an empirical proof of better quality of the new methods. From theoretical point of view, two reasons can be given for which the proposed method F is better than others. Firstly, the proposed method has more knowledge of objects at the classification than the classical *k-NN* method does. This is due to the fact that classification results are used for a number of classifiers with different parameters k. It is generally known that there is no way to determine optimal value of k for all data. For each data, optimal k must be determined through many experiments.

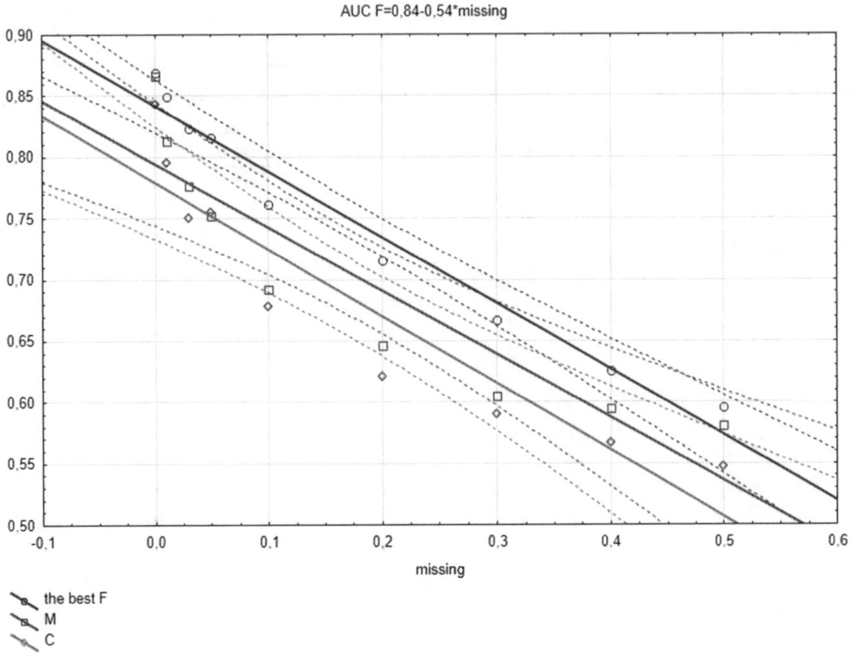

Fig. 4.3 The method F (with the best aggregations) in comparison to the M and C methods

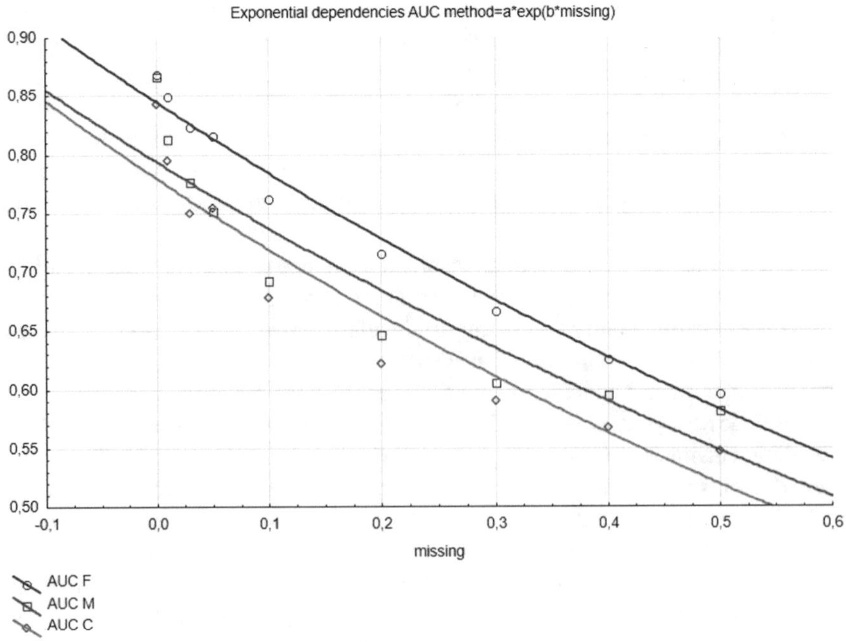

Fig. 4.4 Exponential dependencies for methods

In this method, different k are examined simultaneously, which immediately reveals the knowledge about how the test object is evaluated from the point of view of different k. A separate issue is how to use this knowledge, i.e. how to aggregate it, to extract the most correct (in line with reality) conclusions regarding the decision class to which the test object belongs. We propose two aggregation approaches, namely with and without using interval-valued fuzzy calculus. Both approaches give good results, although the method with interval-valued fuzzy calculus involved (or more generally the interval modeling) is a better one. Secondly, the proposed method uses interval-valued methods to aggregate uncertainty intervals, not the values of membership functions (single numerical values). Therefore, there is more knowledge in this method at the beginning. As the next step, this knowledge is aggregated to make the final decision about the belonging of the test object to a decision class. As a result, the classifier F, with the interval-valued methods involved, has a better chance of making a good decision because it has more knowledge. However, the aggregation must be done in such a way that the quality improvement in the classification is obtained. The methods of aggregation of intervals are based on human intuitions how to aggregate information from various sources (this information is represented in the form of uncertainty intervals). Practical experiments confirmed the usefulness of these intuitions, because the quality of classification has increased.

References

1. Michie, D., Spiegelhalter, D.J., Taylor, D.J.: Machine Learning, Neural and Statistical Classification. Ellis Horwood Limited, England (1994)
2. Pawlak, Z., Skowron, A.: Rudiments of rough sets. Inf. Sci. **177**, 3–27 (2007)
3. Bazan, J.G.: Hierarchical classifiers for complex spatio-temporal concepts. Transactions on Rough Sets IX, pp. 474–750. Springer, Berlin (2008)
4. Bazan, J.G., Buregwa-Czuma, S., Jankowski, A.: A domain knowledge as a tool for improving classifiers. Fundam. Inform. **127**(1–4), 495–511 (2013)
5. Bazan, J.G., Bazan-Socha, S., Buregwa-Czuma, S., Dydo, L., Rzasa, W., Skowron, A.: A classifier based on a decision tree with verifying cuts. Fundam. Inform. **143**(1–2), 1–18 (2016)
6. Buregwa-Czuma, S., Bazan, J.G., Bazan-Socha, S., Rzasa, W., Dydo, L., Skowron, A.: Resolving the conflicts between cuts in a decision tree with verifying cuts (The best application paper award). In: Proceedings of IJCRS 2017, Olsztyn, 3–7 July. Lecture Notes in Computer Science (LNCS), vol. 10314, pp. 403–422. Springer (2017)
7. Bailey, T., Jain, A.: A note on distance-weighted k-nearest neighbor rules. IEEE Trans. Syst. Man, Cybern. **8**, 311–313 (1978)
8. Dudani, S.A.: The distance-weighted k-nearest-neighbor rule. IEEE Trans. Syst. Man, Cybern., SMC **6**, 325–327 (1976)
9. Frank, E., Hall, M.A., Witten, I.H.: The WEKA workbench. Online Appendix for Data Mining Practical Machine Learning Tools and Techniques, 4th edn. Morgan Kaufmann, Burlington (2016)
10. Hall, M., Frank, E., Holmes, G., Pfahringer, B., Reutemann, P., Witten, I.H.: The WEKA data mining software: an update. SIGKDD Explor. **11**(1), 10–18 (2009)
11. Dubois, D., Prade, H.: Gradualness, uncertainty and bipolarity: making sense of fuzzy sets. Fuzzy Sets Syst. **192**, 3–24 (2012)
12. Zadeh, L.A.: The concept of a linguistic variable and its application to approximate reasoning-I. Inf. Sci. **8**, 199–249 (1975)

13. Bentkowska, U.: New types of aggregation functions for interval-valued fuzzy setting and preservation of pos-B and nec-B-transitivity in decision making problems. Inf. Sci. **424**, 385–399 (2018)
14. Dubois, D., Prade, H.: Possibility Theory. Plenum Press, New York (1988)
15. Fawcett, T.: An introduction to ROC analysis. Pattern Recognit. Lett. **27**(8), 861–874 (2006)
16. Swets, J.A.: Measuring the accuracy of diagnostic systems. Science **240**, 1285–1293 (1988)
17. UC Irvine Machine Learning Repository: http://archive.ics.uci.edu/ml/
18. De Waal, T., Pannekoek, J., Scholtus, S.: Handbook of Statistical Data Editing and Imputation, vol. 563. Wiley, Hoboken (2011)
19. Grzymala-Busse, J.W.: Three approaches to missing attribute values: a rough set perspective. Stud. Comput. Intell. (SCI) **118**, 139–152 (2008)
20. Dyczkowski, K.: Intelligent Medical Decision Support System Based on Imperfect Information. The Case of Ovarian Tumor Diagnosis. Studies in Computational Intelligence. Springer, Berlin (2018)
21. Wójtowicz, A., Żywica, P., Stachowiak, A., Dyczkowski, K.: Solving the problem of incomplete data in medical diagnosis via interval modeling. Appl. Soft Comput. **47**, 424–437 (2016)
22. Żywica, P., Dyczkowski, K., Wójtowicz, A., Stachowiak, A., Szubert, S., Moszyński, R.: Development of a fuzzy-driven system for ovarian tumor diagnosis. Biocybern. Biomed. Eng. **36**(4), 632–643 (2016)
23. Żywica, P., Wójtowicz, A., Stachowiak, A., Dyczkowski, K.: Improving medical decisions under incomplete data using intervalvalued fuzzy aggregation. In: Proceedings of the IFSA-EUSFLAT 2015, pp. 577–584. Atlantis Press (2015)
24. Wójtowicz, A., Żywica, P., Szarzyński, K., Moszyński, R., Szubert, S., Dyczkowski, K., Stachowiak, A., Szpurek, D., Wygralak, M.: Dealing with uncertinity in ovarian tumor diagnosis. Modern Approaches in Fuzzy Sets, Intuitionistic Fuzzy Sets, Generalized Nets and Related Topics. Vol. II: Applications, pp. 151–158. SRI PAS, Warszawa (2014)
25. Szubert, S., Wójtowicz, A., Moszyński, R., Żywica, P., Dyczkowski, K., Stachowiak, A., Sajdak, S., Szpurek, D., Alcázar, J.L.: External validation of the IOTA ADNEX model performed by two independent gynecologic centers. Gynecol. Oncol. **142**(3), 490–495 (2016)
26. Stachowiak, A., Dyczkowski, K., Wójtowicz, A., Żywica, P., Wygralak, M.: A bipolar view on medical diagnosis in ovaexpert system. In: Andreasen, T., Christiansen, H., Kacprzyk, J., et al. (eds.) Flexible Query Answering Systems 2015, Proceedings of FQAS 2015, Cracow, Poland, October 26–28, 2015. Advances in Intelligent Systems and Computing, vol. 400, pp. 483–492. Springer International Publishing, Cham, Switzerland (2016)
27. Moszyński, R., Żywica, P., Wójtowicz, A., Szubert, S., Sajdak, S., Stachowiak, A., Dyczkowski, K., Wygralak, M., Szpurek, D.: Menopausal status strongly influences the utility of predictive models in differential diagnosis of ovarian tumors: an external validation of selected diagnostic tools. Ginekol. Pol. **85**(12), 892–899 (2014)
28. Dyczkowski, K., Wójtowicz, A., Żywica, P., Stachowiak, A., Moszyński, R., Szubert, S.: An intelligent system for computer-aided ovarian tumor diagnosis. Intelligent Systems 2014, pp. 335–344. Springer International Publishing, Cham (2015)
29. Fix, E., Hodges, J.L.: discriminatory analysis, aonparametric discrimination: consistency properties. Technical Report 4, USAF School of Aviation Medicine, Randolph Field, Texas (1951)
30. Cover, T.M., Hart, P.E.: Nearest neighbor pattern classification. IEEE Trans. Inform. Theory **13**(1), 21–27 (1967)
31. Bermejo, S., Cabestany, J.: Adaptive soft k-nearest-neighbour classifiers. Pattern Recognit. **33**, 1999–2005 (2000)
32. Jozwik, A.: A learning scheme for a fuzzy k-nn rule. Pattern Recognit. Lett. **1**, 287–289 (1983)
33. Keller, J.M., Gray, M.R., Givens, J.A.: A fuzzy k-nn neighbor algorithm. IEEE Trans. Syst. Man Cybern. SMC **15**(4), 580–585 (1985)
34. Moore, R.E.: Interval Analysis, vol. 4. Prentice-Hall, Englewood Cliffs (1966)
35. Bentkowska, U., Bazan, J.G., Rząsa, W., Zaręba, L.: Application of interval-valued aggregation to optimization problem of *k-NN* classifiers for missing values case. Inf. Sci. (under review)

36. Mansouri, K., Ringsted, T., Ballabio, D., Todeschini, R., Consonni, V.: Quantitative structure - activity relationship models for ready biodegradability of chemicals. J. Chem. Inf. Model. **53**, 867–878 (2013)
37. Wolberg, W.H., Mangasarian, O.L.: Multisurface method of pattern separation for medical diagnosis applied to breast cytology. In: Proceedings of the National Academy of Sciences, vol. 87, pp. 9193–9196. U.S.A., Dec 1990
38. Zhang, K., Fan, W.: Forecasting skewed biased stochastic ozone days: analyses, solutions and beyond. Knowl. Inf. Syst. **14**(3), 299–326 (2008)
39. Antal, B., Hajdu, A.: An ensemble-based system for automatic screening of diabetic retinopathy. Knowl. Based Syst. **60**, 20–27 (2014)
40. Japkowicz, N., Shah, M.: Evaluating Learning Algorithms: A Classification Perspective. Cambridge University Press, New York (2011)

Chapter 5
Optimization Problem of k-*NN* Classifier in DNA Microarray Methods

A functional biological clock has three components: input from the outside world to set the clock, the timekeeping mechanism itself, and genetic machinery that allows the clock to regulate expression of a variety of genes.

Jeffrey C. Hall

The microarrays are a particularly interesting tool in modern molecular biology, not only because of the wide spectrum of applications such as analysis of the genome structure, profile gene expression, genotyping, sequencing, but also due to the possibility of testing a large number of objects in one experiment (cf. [1, 2]). However, the identification of relevant information from a huge amount of data obtained using microarrays requires the use of sophisticated bioinformatic methods. Clustering methods or machine learning algorithms are applied. However, when such methods are used, there is a problem of lowering their performance on test data due to the large number of attributes (columns). In this chapter microarray methods are applied for identification of marker genes. Our aim is to show that the quality of classification in the case of large number of attributes may be improved while using the microarray methods and interval modeling. There will be considered the so called *vertical decomposition* of a table representing a given data set.

U. Bentkowska, *Interval-Valued Methods in Classifications and Decisions*, Studies in Fuzziness and Soft Computing 378,
https://doi.org/10.1007/978-3-030-12927-9_5

5.1 DNA Microarray Methods from Biological Point of View

In contemporary science, generally recognized as true is the so-called *Central Dogma of Molecular Biology*, which states that genetic information, which in living organisms is stored in the DNA sequence, determines the functioning of organisms. The flow of genetic information takes place from DNA through RNA to protein. Knowledge of this fact facilitates a better understanding of cell function through various analysis of the structure of DNA, RNA and proteins that perform the main cellular functions. One of the methods of such analysis is the study of gene expression, understood as a process in which the genetic information contained in a gene is read and written down to its products, which are proteins or different forms of RNA. The intensity (level) of the expression of individual genes reflects the functioning of the cell (it reflects the state of the processes in the cell). *DNA microarrays* are used to measure the level of expression of genes and moreover they measure the level of expression of a huge number of genes simultaneously.

There exist diverse methods of sample labelling and the principles of construction and application of microarrays. These are high density oligonucleotide and low density microarrays. They may be applied for example for identification of bioterrorism agents, as well as water and food contaminating bacteria and other selected microorganisms, and also antibiotic resistance testing (cf. [3–8]). Due to the fact that the microarray method allows to identify thousands of genes in one experiment, this method opens new perspectives in areas such as epidemiological studies, determination of sources of disease outbreaks, detection of new genotypes and subtypes, or examining of the geographical spread of the biological agents. To solve the given practical problems there are implemented higher analysis algorithms such as pattern search and machine learning approaches [9].

5.2 DNA Microarray Methods from Information Technology Point of View

A DNA microarray (or DNA chip) is a glass or plastic plate with microscopic-sized fields plotted at regular positions, containing DNA fragments that differ from each other. These fragments are probes that detect complementary DNA or RNA molecules. On the surface of a few square centimeters, in micrometer distances there are placed probes that allow to study the expression of even tens of thousands of genes simultaneously. The level of expression is read by the detection of laser light passing through a sample, i.e. a small piece of biological material from a living organism. This level is represented by a rational number. The main goals of microarray experiments are: discovering and identifying genes or groups of genes with similar or different expression and using expression for the determination (classification) of biological

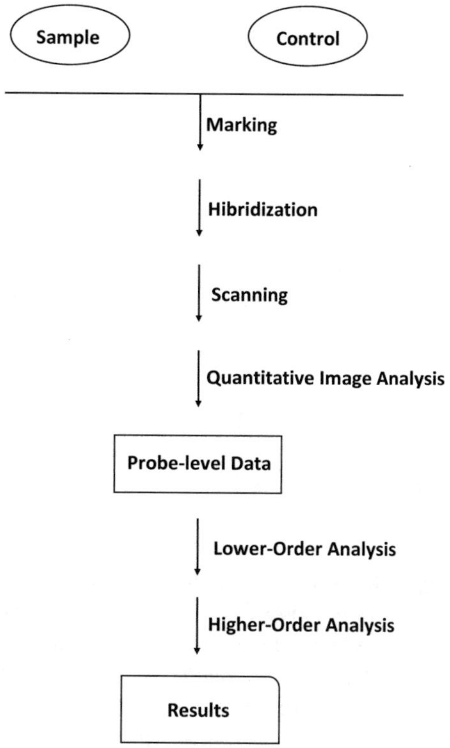

Fig. 5.1 Schematic description of microarray experiment

samples. The main stages of the microarray experiment are presented in Fig. 5.1 (cf. [9]).

The obtained probe-level data are normalized, first locally on a single probe and then globally on all microarrays involved in the experiment. At each stage, the quality of microarrays is checked, excluding those that have serious technical defects. The next stage is called gene filtering and aims to select the so-called differentiating genes, the expression of which varies significantly in the subjects conditions, and the rejection of those that give no signal. On the basis of such a reduced set of genes, higher-order analyzes are conducted, looking for groups of genes with a similar / different expression profile. At the end the results obtained are subjected to biological interpretation, consisting in linking the observed changes in the level of gene expression with physiological or pathological processes occurring in the examined organism. Schematic description of the data analysis process is presented in Fig. 5.2 (cf. [9]).

Higher-order analysis focuses on the most important from biological point of view problem, namely the search for biological dependencies in the data obtained from the microarray experiment. The simplest and the most-prone method of searching for trends in the microarray data is to combine genes and samples into groups with a common profile, the so-called clusters. The basis of operation of each grouping

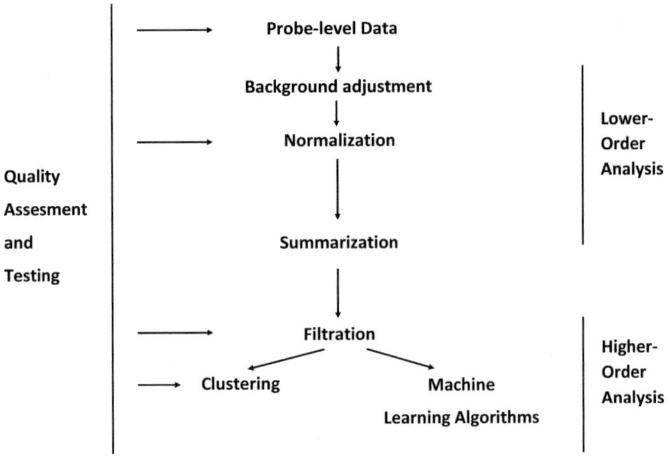

Fig. 5.2 Schematic description of the data analysis process

algorithm is to create a dissimilarity matrix according to which it will be possible to organize elements from the most to the least similar to each other. Different types of gene grouping methods are most often used in the research mechanisms responsible for the regulation of gene expression, metabolic pathways, stress responses or therapy. A completely different approach requires experiments designed to identify the so-called marker genes used in medical diagnosis. In such situations, learning algorithms are usually used. Most experiments are focused on determining the best method for diagnosing a given type of illness.

So far, many algorithms for classifying samples have been proposed. The simplest of them use combinations of linear dependencies LDA (linear discriminant analysis), linear regression modifications (logistic regression) or weighted voting (cf. [9]). Among the more complex algorithms there are, for example, those that are based on a network of cooperating, layered elements that perform classification calculations, these are the so-called neural networks. These include, for example, the multilayer perceptrons algorithm MLPs or the Bayesian strategy based PNN (probabilistic neural network). An interesting and quite popular algorithm based on the so-called support vector machines SVMs, instead of elaborating complicated functions in space equal to the number of genes in training samples (as artificial neural networks do) maps the introduced elements to a hypothetical space with a larger number of dimensions, and then separates them using simple linear classification functions.

In our experiments we applied classification methods for identification of marker genes that are based on the *k-NN* algorithm. The *k-NN* has already proved to be effective for successful classification of microarrays (cf. [10]). It has also been shown that reduction of dimension can improve the performance of *k-NN* for this task (cf. [11, 12]). From Information Technology point of view, the microarray is represented as a rectangular array of rational numbers, where the rows of the table correspond to the genome and the columns to the biological samples. On the other hand, the values in the table correspond to the value of the gene expression from a given row measured for material from a given column. However, for Information Technology experiments, the above table is often transposed to obtain another table, which we call a *microarray decision board*, where the rows correspond to samples (for example patients), and the columns to genome. In addition, the last attribute is added, the so-called decisive one, describing the property of patients requiring prediction based on the value of gene expression. This allows, for example, prediction of the fact whether a given patient is immune to a certain disease, what is the level of the patient's risk during the illness, which medicine should be given to the patient, whether the patient can be cured of a certain disease, etc. Microarray decision boards are characterized by a large number of columns (even over 60,000) and a small number of rows (up to several hundred). This disproportion results from the large number of known genes and the large financial cost of the analysis of each biological sample. It is easy to see that new methods have to be developed to explore the microarray data, as previous methods usually cannot work for such a large number of attributes. This problem appears in the construction of classifiers that we are considering in this chapter. For example, the typical *k-NN* method, using the Euclidean distance, must investigate the diversity of the expression of object pairs on all genes, which becomes ineffective in time. In addition, the information represented by the attributes is very redundant, namely the attributes strongly depend on each other. Therefore, the simple calculation of the Euclidean distance between the two objects does not reflect their real diversity. The existing methods of dealing with this problem rely on the selection or grouping of attributes. The selection of attributes consists in selecting a relatively small number of attributes and based on the classifier induction on this set. The criteria for selecting attributes can be different. One of them is the approach based on the so-called *positive area*. Although this approach allows to obtain effective classifiers, they do not use all the knowledge gathered in the decision table, and thus usually do not allow to obtain high quality classifications. The grouping of attributes is based on the fact that the attributes are divided into groups for which classifiers are constructed separately. In the end, the operation of these classifiers must be properly aggregated by a special meta-classifier. This method will be presented in this chapter. General scheme of such procedure is presented in Fig. 5.3.

In situations where the classifier construction method for microarray data is used, based on grouping attributes, constructing classifiers for groups, and using the meta aggregation classifier, the following problems should be solved. How should we obtain groups of attributes for which the classifiers will be generated? How to generate classifiers for these groups of attributes? How to construct a meta classifier aggregating the results of previously constructed classifiers?

Here we present a random method to extract attribute groups, which consists in constructing a group of attributes with a non-return drawing, where the stop condition of such a group generation is based on the assumption that the group construction ends when the *positive region* of the selected attribute group with respect to the decision attribute will reach 100%, i.e. each two objects from a subtable of different decision classes are distinguishable by means of at least one attribute from the selected attribute

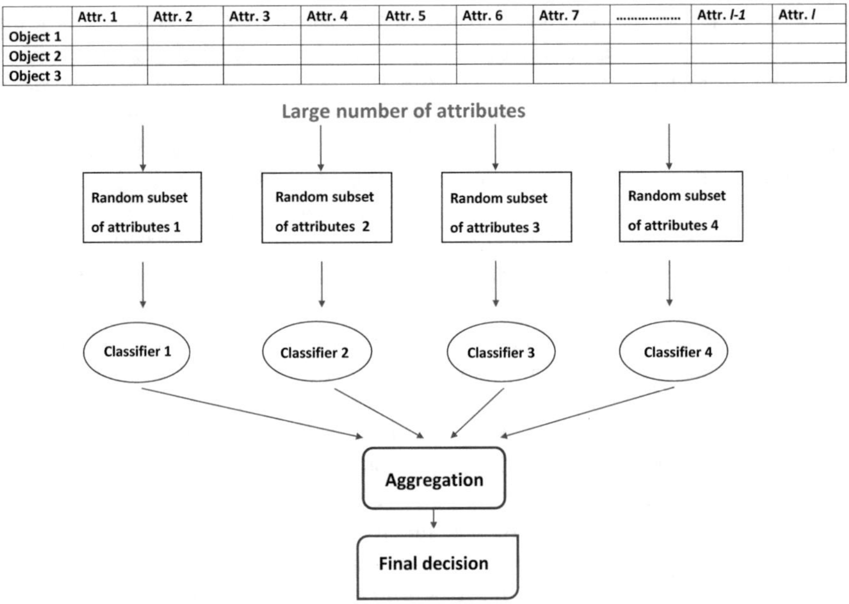

Fig. 5.3 Vertical decomposition

group. We assume that the number of selected attribute groups, denoted by s, can range from a dozen to a few hundred. For selected attribute groups, specifically for subtables composed of selected attributes and a decision attribute, *k-NN* classifiers are used with the fixed k (for example $k = 3, 5, 7, 9$). The meta classifier, that resolves conflicts between the lower level classifiers, is constructed in two ways. The first method (called the M method, cf. Algorithm 5) consists in calculating the arithmetic mean of the weights (i.e. classification coefficients of belonging to the main class) determined for the test object by the lower level classifiers and checking whether this average is above or below a fixed threshold (the method of determining thresholds is described in Sect. 5.3). The second method (called the F method) uses the aggregation of uncertainty intervals calculated on the basis of the lower level classifiers (cf. Sect. 5.3).

If it comes to the classifier M, a test object may be classified using the *k-NN* method with diverse parameters k. In the performed experiments we used the parameters $k = 3, 5, 7, 9$. In each of these cases a different set of neighbors is used, which can classify the test object with different levels of confidence for different decision classes. This is due, among other things, to the fact that various objects appear in particular environments. In Algorithm M for a fixed k there is used aggregation (by the arithmetic mean) of weights obtained by the *k-NN* classifier applied to the given number of subtables. This is similar method to the one presented in Algorithm M (cf. Algorithm 3 in Chap. 4) but in that method certainty coefficients (weights) obtained

Algorithm 5: Classification of a tested object by the M classifier for DNA microarray data

Input:
1. training data set represented by decision table $\mathbf{T} = (U, A, d)$, where $n = card(U)$ and $l = card(A)$,
2. *k-NN* classifier, e.g., $k \in \{3, 5, 7, 9\}$,
3. parameter s is a natural number greater than 1, e.g., $s = 50$,
4. a parameter $b \in (0.0, 1.0)$ and a threshold $t \in (0.0, 1.0)$ determined experimentally based on the DACC measure,
5. test object u.

Output: The membership of the object u to the "main class" or to the "subordinate class"

begin
  Choose randomly s subtables from $\mathbf{T} = (U, A, d)$ (any subtable is associate with a group of attributes selected from A and all objects from U), in such a way that the generation of group is finished when the positive region of selected attributes with respect to the decision attribute reaches 100%;
  for $i := 1$ **to** s **do**
    Compute certainty coefficient ("main class" membership probability) for the given test object u using the classifier *k-NN* and assign it to p_i
  end
  Determine the final certainty coefficient p for the object u, where $p = b \cdot \frac{p_1 + \cdots + p_s}{s}$
  if $p > t$ **then**
    return *u belongs to the "main class"*
  else
    return *u belongs to the "subordinate class"*
  end
end

for individual *k-NN* classifiers were aggregated (not the fix classifier *k-NN* applied for a given number of subtables).

5.3 A Method of Constructing a Complex Classifier

The proposed method consists in using a special procedure that is a complex classifier. At the beginning, we set a certain decision class, which we call the main class. We assume that there is a second decisional class in the data set, which we call a subordinate class. The main class can be, for example, the class of patients with a given disease, and the subordinate class consists of healthy patients.

Next, we generate with the help of a random draw without returning s attribute groups so that the positive area of each of these subtable groups with respect to the original decision attribute is 100%. The positive area of a given G group of the conditional attributes with respect to a given decision attribute, is the number $(\tilde{n}/n) \cdot 100\%$, where n is the number of all objects (rows) in the decision table. Moreover, $\tilde{n}$ is the number of such objects that there is no other object for any of these objects that has the same values on all attributes in G, but with a different value of the decision attribute. For each of these subtables, we construct the *k-NN* classifiers for a fixed k. In the case of the M method, we calculate the arithmetic mean of the weights determined for the test object by s lower level classifiers and we check if this average is above or below a fixed threshold t (for example $t = 0.5$). The threshold t is selected experimentally (it will be explained later).

In the case of the method F (cf. Algorithm 6), the subtables are divided into h groups ($i = 1, \ldots, h$), in each such group i there are v classifiers $G_1^i, \ldots, G_v^i$ ($s = h \cdot v$). Based on each group for the test object an uncertainty interval for this object is calculated $[\min_i, \max_i]$, where $\min_i$ is the minimum of all weights calculated for the test object by the classifiers from the given group. Similarly, $\max_i$ is the maximum of all weights calculated for the test object by classifiers from the given group. After determining the uncertainty intervals, they are aggregated using one of the aggregation operators. In this way we obtain an aggregated uncertainty interval, which is used to classify the test object in the following manner. The arithmetic mean of the upper and lower end of the aggregated uncertainty interval is calculated and then it is checked whether this average is above or below a fixed threshold t (for example $t = 0.5$). The threshold t is selected experimentally (it will be explained later).

It is important to note that simple aggregation using the arithmetic mean does not use special aggregation methods that are available in the F methods (with diverse interval-valued aggregations applied). These methods can lead to classification improvement because they take into account additional information about the minimum and maximum classification weight in the individual classifiers and then they aggregate this information in the adequate way.

In contrast to Chap. 4, here we do not use the AUC parameter to assess the quality of the analyzed methods. AUC value (which is the area under the ROC curve) has its advantages which were recalled in the mentioned Chap. 4. However, providing the AUC parameter, we lose information about specific points at the ROC curve. Sometimes we are only interested in some specific points. For example, in medical applications, where we need to determine the membership of a test object to a certain decision class, related to a patient's medical condition, we may want to find a point on the ROC curve that meets certain specific conditions. In this chapter, we assume that we are looking for such a point on the ROC curve, for which the following measure $DACC = ACC \cdot (1 - |SN1 - SN2|)$ takes the maximum value (where DACC comes from *accuracy differentiating the decision classes or concepts*). ACC is the accuracy value for the whole data set and SN1 and SN2 are the sensitivity values for the first and the second decision class, respectively. Accuracy is a measure of the ratio of the number of correctly classified objects to all evaluated objects, namely

$$ACC = \frac{TP + TN}{TP + TN + FP + FN}.$$

It is easy to see that the above DACC measure basically relies on global accuracy, but the difference in sensitivity of both decision classes lowers the value of this measure. Therefore, the higher the difference between the sensitivity for both classes, the lower the value of the DACC measure. In other words, we are looking for a point on the ROC curve corresponding to the classification method that achieves high global accuracy, but the sensitivity for both decision classes is similar. Therefore, the point is to avoid the situation that we consider the method to be good when it has high accuracy, but it results from the high sensitivity for one of the decision classes, whereas the sensitivity for the second class is low. This is particularly important when the data is unbalanced, i.e. one decision class dominates with the number over the other. We encounter this situation in some of the microarray data analyzed in the presented experiment. Moreover, in order to avoid the negative effects of unbalanced decision classes in the analyzed data, an additional mechanism for adjusting the value of the DACC measure was used in the experiments. Namely, the weight values provided by the classification of test objects by the lower level classifiers were multiplied by a special, determined parameter called *the multiplier b*, whose aim was to offset too much impact on the classification from the larger decision class. In the case of the M method, one weight value provided by the lower level classifiers was multiplied by b. However, in the case of the F method, both ends of the interval were multiplied, but each of them being multiplied by another multiplier value, provided by the lower level classifiers. All multipliers were selected experimentally, to each data set separately, to select the point of the ROC curve that maximizes the value of the DACC quality measure. The above-mentioned threshold t, used to classify the test object to the main or subordinate class, was selected in the analogous way.

Now, we present a brief analysis of the computational time complexity of the Algorithm 6. Let assume that the classical *k-NN* algorithm works in time of order $O(n \cdot l)$, where n is the number of objects in the set U, and l is the number of attributes in the set A. Thus time complexity of the Algorithm 6 is of order $O(s \cdot n \cdot f)$, where s is the number of subtables, n is the number of objects in the set U, and f is the maximum number of conditional attributes, which, at random selection of attributes, allows to obtain the positive region of selected attributes with respect to the decision attribute equal 100%. Certainly, in practice $f << l$ (f is much smaller than l). Therefore, it easy to see that this complexity is not high and the algorithm can be used in practical applications for exploration of DNA microarray data sets, i.e. in the case when the classical *k-NN* algorithm (with the use of all conditional attributes) cannot be used.

5.4 Details of Experiments

The experiments (cf. [13]) were performed with the use of WEKA API library (cf. [14, 15]). Table 5.1 presents the general characteristics of the data sets that were used in experiments. They come from ELVIRA Biomedical Data Set Repository [16].

We used data sets that concern microarrays made on biological material from patients with diseases such as:

- Colon (Colon Tumor, cf. [17]) - contains 62 samples collected from colon-cancer patients; among them, 40 tumor biopsies are from tumors (labelled as 'negative') and 22 normal (labelled as 'positive') biopsies are from healthy parts of the colons of the same patients; two thousand out of around 6,500 genes were selected based on the confidence in the measured expression levels.

Algorithm 6: Classification of a tested object by the F classifier for DNA microarray data

---

Input:
1. data set represented by decision table $\mathbf{T} = (U, A, d)$, where $n = card(U)$ and $l = card(A)$,
2. *k-NN* classifier, e.g. $k \in \{3, 5, 7, 9\}$,
3. parameters s, h, v are natural numbers, where $s = h \cdot v$, e.g., $h = 10$ and $v = 50$,
4. aggregation function $\mathcal{A}$,
5. parameters $b_1, b_2, \in (-1.0, 0.0) \cup (0.0, 1.0)$ and a threshold $t \in (0.0, 1.0)$ determined experimentally based on the DACC measure,
6. test object u

Output: The membership of the object u to the "main class" or to the "subordinate class"

1 **begin**
2   Choose randomly s subtables from $\mathbf{T} = (U, A, d)$ (any subtable is associate with a group of attributes selected from A and all objects from U), in such a way that the generation of group is finished when the positive region of selected attributes with respect to the decision attribute reaches 100%;
3   Divide s subtables into h groups, where in each group there are $G_1^i, \ldots, G_v^i$ classifiers, where $i = 1, \ldots, h$ $(s = h \cdot v)$;
4   **for** $i := 1$ **to** h **do**
5   **for** $j := 1$ **to** v **do**
6   Compute certainty coefficient for *k-NN* classifier G_j^i and assign this value to p_j^i;
7   **end**
8   Compute $min\{p_1^i, \ldots, p_v^i\}$ and assign it to min_i;
9   **if** $b_1 > 0.0$ **then**
10           $min_i := b_1 \cdot min_1$ (decreasing the value of min_i)
11   **else**
12           $min_i := 1.0 - ((-1) \cdot b_1) \cdot (1.0 - min_i)$ (increasing the value of min_i)
13   **end**
14   Compute $max\{p_1^i, \ldots, p_v^i\}$ and assign it to max_i;
15   **if** $b_2 > 0.0$ **then**
16           $max_i := b_2 \cdot max_i$ (decreasing the value of max_i)
17   **else**
18           $max_i := 1.0 - ((-1) \cdot b_2) \cdot (1.0 - max_i)$ (increasing the value of max_i)
19   **end**
20   **end**
21   Determine the uncertainty interval $[down(u), up(u)]$ for the object u by aggregating (with the use of interval-valued aggregation function $\mathcal{A}$) the intervals $[min_1, max_1], \ldots, [min_h, max_h]$;
22   Determine the final certainty coefficient $\mathbf{p} = \frac{down(u)+up(u)}{2}$ for the object u;
23   **if** $\mathbf{p} > t$ **then**
24   **return** *u belongs to the "main class"*;
25   **else**
26   **return** *u belongs to the "subordinate class"*;
27   **end**
28 **end**

Table 5.1 Characteristics of the microarray data sets (ELVIRA Biomedical Dataset Repository)

Data set	Objects	Attributes	Classes
Colon	62	2,000	2
Leukemia	72	7,129	2
Lymphoma	47	4,026	2
Lung cancer	181	12,533	2
Ovarian cancer	253	15,154	2
Prostate	136	12,600	2

- Leukemia (Acute lymphocytic leukemia - ALL and Acute myeloid leukemia - AML, cf. [18]) - training dataset consists of 38 bone marrow samples (27 ALL and 11 AML); over 7,129 probes from 6,817 human genes, moreover 34 samples testing data is provided, with 20 ALL and 14 AML.
- Lymphoma (cf. [19]) - there are considered distinct types of diffuse large B-cell lymphoma (DLBCL) using gene expression data; there are 47 samples, 24 of them are from 'germinal centre B-like' group while 23 are 'activated B-like' group, each sample is described by 4,026 genes.
- Lung cancer (cf. [20]) - deals with classification between malignant pleural mesothelioma - MPM and adenocarcinoma - ADCA of the lung; there are 181 tissue samples (31 MPM and 150 ADCA), the training set contains 32 of them, 16 MPM and 16 ADCA, the rest 149 samples are used for testing, each sample is described by 12,533 genes.
- Ovarian cancer (cf. [21]) - identifies proteomic patterns in serum that distinguish ovarian cancer from non-cancer (this study is significant to women who have a high risk of ovarian cancer due to family or personal history of cancer); the proteomic spectra were generated by mass spectroscopy and the data set provided here is 6-19-02, which includes 91 controls (normal) and 162 ovarian cancers; the raw spectral data of each sample contains the relative amplitude of the intensity at each molecular mass / charge (M/Z) identity; there are total 15,154 M/Z identities; the intensity values were normalized according to the formula: NV = (V-Min)/(Max-Min), where NV is the normalized value, V the raw value, Min the minimum intensity and Max the maximum intensity; the normalization is done over all the 253 samples for all 15,154 M/Z identities (after the normalization, each intensity value is to fall within the range of 0–1).
- Prostate (Prostate cancer, cf. [10]) - provides tumor versus normal classification; there are two sources of data; training set from the first source contains 52 prostate tumor samples and 50 non-tumor (labelled as 'normal') prostate samples with around 12,600 genes; an independent set of testing samples from the second source was also prepared, which is from a different experiment and has a nearly 10-fold difference in overall microarray intensity from the training data; besides, extra genes contained in the testing samples were removed.

Below there are presented details of the performed experiments (the M method and the F method).

- Due to the small size of the data, the experiments were carried out using the leave-one-out cross validation method. This method is based on the fact that each of the table objects is classified in such a way that all other objects form a training table for which a classifier is designed to classify only this one object.
- In the experiments, aggregation operators (4.1)–(4.6) are used where aggregation operator (4.4) is applied with $p = 3$.
- Experiments are carried out for the following classifiers:
  - the classifier M - uses aggregation of certainty coefficients by the arithmetic mean, completely without using the uncertainty intervals cf. Algorithm 5 (k-NN classifiers for $k = 3, 5, 7, 9$ are used).
  - the classifier F - it is the algorithm for aggregating uncertainty intervals (k-NN classifiers for $k = 3, 5, 7, 9$ are used) described in Sect. 5.3 (cf. Algorithm 6).
- The final confidence interval after the aggregation method is used in such a way that the arithmetic mean of the lower and upper ends is calculated and this average is returned as the final classification result to the main class (the threshold t is taken into account).
- Each experiment is repeated 20 times and the average ACC and standard deviation are reported.
- In each experiment, the main class has always been determined, which is given as an experiment parameter. We assume that all analyzed data have only two decision classes.
- One of the main parameters of the experiment is the number of attribute groups, and also the number of subtables created on these groups and lower level classifiers built for these subtables. These parameters for the F method are divided into h groups, each with v tables.
- The Euclidean distance was used for k-NN classical algorithms in WEKA API library.

In Tables 8.4, 8.5 and 8.6 some of the results of the performed experiments are presented. In the next section we provide the analysis of the obtained results.

5.5 Discussion and Statistical Analysis of the Results

When we analyze the results in Tables 8.4, 8.5 and 8.6 (for the remaining data sets the results are analogous), we see that generally the highest ACC values were obtained for the greatest number of subtables s. The value $s = v \cdot h$ (cf. Sect. 5.3, Algorithm 6) depends on the value h - the number of aggregated intervals (i.e. the source of information, in the performed experiments $h \in \{10, 5, 3\}$) and the value v - the number of lower level classifiers in each group used to determine a given uncertainty interval ($v \in \{50, 10, 5\}$). Because of the large amount of data, in Tables 8.4, 8.5 and 8.6, we present the results only for the value $h = 10$. The analysis was performed with the

Table 5.2 The p value for comparison of the highest ACC values for the F and M methods

NoClassVert	Colon	Leukemia	Lymphoma
50	0.0000	0.0000	0.0000
10	0.0000	0.0000	0.0000
5	0.0004	0.0001	0.0001

use of Statistica program. In Table 5.2 there are presented the results of the statistical analysis of the obtained average ACC values (for 20 repetitions of experiments). It was applied average comparison test with respect to the obtained average values of ACC. We see that for each number v of the lower level classifiers (in Table 5.2 this value is denoted by NoClassVert) the highest value for the method F was always statistically better than the highest value for the method M.

Similarly to the results presented in Chap. 4 we conclude that application of the proposed in this chapter classification method F (with the interval-valued aggregation operators involved) always gave better classification results than applying aggregation with the use of the arithmetic mean. This is due to the fact that simple aggregation using the arithmetic mean does not use special aggregation methods that are available in the F methods (with diverse interval-valued aggregations applied). These methods can lead to classification improvement because they take into account additional information about the minimum and maximum classification weight in individual classifiers and then they aggregate this information in adequate way. Moreover, analogously to Chap. 4, diverse interval-valued aggregation operators were the winning ones both with respect to the number of lower level classifiers v and the number of aggregated intervals h. Among considered aggregation operators there are recently introduced pos-aggregation functions and nec-aggregation functions (discussed in the first part of this monograph). As a result we see that they proved to be useful in practice. Finally, we proved that the complexity of the considered Algorithm 6 is not high and this algorithm can be used in practical applications for exploration of DNA microarray data sets, i.e. in the case when the classical *k-NN* algorithm (with the use of all conditional attributes) cannot be used.

Ending this chapter, we want to stress that in DNA microarray experiments it is important to confront the obtained results with the information included in literature and databases, including Gene Ontology. The use of all kinds of statistical tests assessing the credibility of the hypotheses is of great importance, especially due to the specificity of the data microarrays, in which the number of observed features (genes) many times exceeds the number of samples (unit observations). Moreover by changing a bit the statistical parameter selection, for example reducing the selectivity of the applied methods, there is a possibility of obtaining additional, biologically relevant information (cf. [9]). Moreover, as it was pointed out in [12] none of the methods for dimension reduction applied for microarray and *k-NN* classification used in that study consistently gave the best accuracy on all data sets. This is why for the future research it would be useful to consider different dimensionality reduction methods along with the interval modeling (including aggregation operators) which proved to be effective in the experiments described in this chapter.

References

1. Bodrossy, L.: Diagnostic oligonucleotide microarrays for microbiology. In: Blalock, E. (ed.) A Beginners Guide to Microarrays, pp. 43–92. Kluwer Academic Publisher, New York (2003)
2. Heller, M.J.: DNA microarray technology: devices, systems, and applications. Annu. Rev. Biomed. Eng. **4**, 129–153 (2002)
3. Fukushima, M., Kakinuma, K., Hayashi, H., et al.: Detection and Identification of mycobacterium species isolates by DNA microarray. J. Clin. Microbiol. **41**, 2605–2615 (2003)
4. Karczmarczyk, M., Bartoszcze, M.: DNA microarrays - new tool in identification of bilogical agents. Przegl. Epidemiol. (In Pol.) **60**, 803–811 (2006)
5. Robertson, B.H., Nicholson, J.K.A.: New microbiology tools for public health and their implications. Annu. Rev. Public Health **26**, 281–302 (2005)
6. Stenger, D.A., Andreadis, J.D., Vora, G.J., et al.: Potential applications of DNA microarrays in biodefenserelated diagnostics. Curr. Opin. Biotechnol. **13**, 208–212 (2002)
7. Straub, T.M., Quinonez-Diaz, M.D., Valdez, C.O., et al.: Using DNA microarrays to detect multiple pathogen threats in water. Water Supply **2**, 107–114 (2004)
8. Yu, X., Susa, M., Knabbe, C., et al.: Development and Validation of a Diagnostic DNA Microarray To Detect Quinolone-Resistant Escherichia coli among Clinical Isolates. J. Clin. Microbiol. **42**, 4083–4091 (2004)
9. Stępniak, P., Handschuh, L., Figlerowicz, M.: DNA microarray data analysis (in Polish). Biotechnologia **4**(83), 68–87 (2008)
10. Singh, D., et al.: Gene expression correlates of clinical prostate cancer behavior. Cancer Cell **1**, 203–209 (2002)
11. Deegalla, S., Boström, H.: Reducing high-dimensional data by principal component analysis vs. random projection for nearest neighbor classification. In: Proceedings of the 5th International Conference on Machine Learning and Applications, ICMLA 2006, pp. 245–250. IEEE Computer Society, Washington, DC, USA (2006)
12. Deegalla, S., Boström, H.: Classification of microarrays with kNN: comparison of dimensionality reduction methods. In: Yin, H., et al. (eds.) IDEAL 2007, LNCS 4881, pp. 800–809. Springer, Berlin (2007)
13. Bentkowska, U., Bazan, J. G., Rząsa, W., Zaręba, L.: Application of interval-valued aggregation to optimization problem of *k-NN* classifiers in DNA microarray methods (under preparation)
14. Frank, E., Hall, M.A., Witten, I.H.: The WEKA Workbench. Online Appendix for Data Mining Practical Machine Learning Tools and Techniques, 4th edn. Morgan Kaufmann, Burlington (2016)
15. Hall, M., Frank, E., Holmes, G., Pfahringer, B., Reutemann, P., Witten, I.H.: The WEKA data mining software: an update. SIGKDD Explor. **11**(1), 10–18 (2009)
16. ELVIRA Biomedical data set repository. http://leo.ugr.es/elvira/DBCRepository/
17. Alon, U., et al.: Broad patterns of gene expression revealed by clustering analysis of tumor and normal colon tissues probed by oligonucleotide arrays. PNAS **96**, 6745–6750 (1999)
18. Golub, T., Slonim, D., Tamayo, P., Huard, C., Gaasenbeek, M., Mesirov, J., Coller, H., Loh, M., Downing, J., Caligiuri, M., Bloomfield, C., Lander. E.: Molecular classification of cancer. Class discovery and class prediction by gene expression monitoring. Science **286**(5439), 531–537 (1999)
19. Alizadeh, A., Eisen, M., Davis, R., Ma, C., et al.: Distinct types of diffuse large B-cell lymphoma identified by gene expression profiling. Nature **403**, 503–511 (2000)
20. Gordon, G., Jensen, R., Hsiao, L., Gullans, S., Blumenstock, J., Ramaswamy, S., Richards, W., Sugarbaker, D., Bueno, R.: Translation of microarray data into clinically relevant cancer diagnostic tests using gene expression ratios in lung cancer and mesothelioma. Cancer Res. **62**(17), 4963–4967 (2002)
21. Petricoin III, E.F., et al.: Use of proteomic patterns in serum to identify ovarian cancer. Lancet **359**, 572–577 (2002)

Chapter 6
Interval-Valued Methods in Medical Decision Support Systems

There is a lot of work out there to take people out of the loop in things like medical diagnosis. But if you are taking humans out of the loop, you are in danger of ending up with a very cold form of AI that really has no sense of human interest, human emotions, or human values.

Louis B. Rosenberg

In this chapter we present the application of methods based on interval modeling and aggregation in OvaExpert computer support system [1] designed for ovarian tumor diagnosis (however applicable also in other medical fields). It was shown that such methods made it possible to reduce the negative impact of lack of data and lead to meaningful and accurate decisions [2–10]. Here the behavior of some new interval-valued operators in OvaExpert is shown, namely there are considered possible and necessary aggregation functions and aggregation functions with respect to admissible linear orders. These aggregation operators were not previously considered in OvaExpert. The results prove that these new aggregation operators may be competitive with others, especially if it comes to the cost matrix results.

The motivation to develop such system was to support doctors in diagnosis of ovarian tumor (this type of cancer is particularly difficult to diagnose and mortality rates have remained high for many years). The system is managing incompleteness of data by interval modeling. It is also integrating existing diagnostic methods to make them usable with incomplete data. It enables to make accurate and high-quality decisions under incomplete information and uncertainty. For the evaluation process in [8] six diagnostic models were selected: two scoring systems - SM and Alc [11, 12] and four regression models - LR1, LR2, Tim and RMI [13–15]. OvaExpert uses interval modeling of incomplete data which enables the uncertaintification of the classical mentioned models. The classical models were created by individual research units, such as the Alcazar model and SM, other methods (like LR1) by organizations (incorporating a number of research centers), such as IOTA (The International

U. Bentkowska, *Interval-Valued Methods in Classifications and Decisions*, Studies in Fuzziness and Soft Computing 378,
https://doi.org/10.1007/978-3-030-12927-9_6

Ovarian Tumor Analysis group which was founded in 1999 by Dirk Timmerman, Lil Valentin and Tom Bourne). The majority are scoring models and models based on logistic regression. In decision module of diagnostic models aggregation of the existing models may be applied in order to take advantage of the synergy of data. Moreover, aggregation methods applied to diverse structures and problems proved to be effective (cf. [16–18]).

6.1 OEA Module of OvaExpert Diagnosis Support System

One of the modules in OvaExpert is OEA. In OEA (cf. [3, 8]) the main approach deployed in the system is based on Ordered Weighted Averaging operation (OWA). In ovarian tumor diagnosis the problem of missing data is commonly encountered. The results presented in [10] confirmed that methods based on interval modeling and aggregation make it possible to reduce the negative impact of lack of data and lead to meaningful and accurate decisions. A diagnostic model developed in this way proved to be better than classical diagnostic models for ovarian tumor. OEA is based on the binary classifier (giving the result *malignant* or *benign*). In OvaExpert [3] there are applied other modules where also multi-class classification is possible. In OEA the aim of the training phase was to optimize the parameters of the aggregation operators and thresholding strategies on different simulated percentages of missing features. In the testing phase, the optimized aggregation operators and thresholding strategies were examined on the test set. This step checked the performance of these aggregation operators on data with the actual missing values. Although interval-valued functions representing models were considered, two possible modes of aggregation were applied. The first, called numerical, uses a single value that represents the whole interval (e.g. the interval's center, lower bound or upper bound). The interval mode, on the other hand, utilizes the whole of the interval information. The study group consisted of 388 patients diagnosed and treated for ovarian tumor in the Division of Gynecological Surgery, Poznan University of Medical Sciences, between 2005 and 2015. The scheme of OEA module is presented in Fig. 6.1 (the source of Fig. 6.1 is the paper [8]).

As it was stated before, application of aggregation methods in medical diagnosis support proved to be fruitful and improved the ability to obtain the final diagnosis (cf. [7]). The presented here results show also the performance of diverse relations for interval comparing, i.e. $\preceq_\pi$, $\preceq_\nu$ and linear orders $\leq_{L'}$. We used analogous methods to the ones presented in [14] and applied our operators in OEA designed for ovarian tumor diagnosis [3]. However, in the training phase (due to the ability of the applied equipment) we performed less repetitions of the evaluation. For some of the examples of aggregation functions we obtained comparable results and for the others even better results, i.e. these examples of aggregation functions proved to be comparable statistically but with lower cost of prediction. We have tested several examples of each considered class of aggregations and diverse thresholding strategies (cf. [10]). We have considered three classes of aggregation functions. Finally, we have chosen

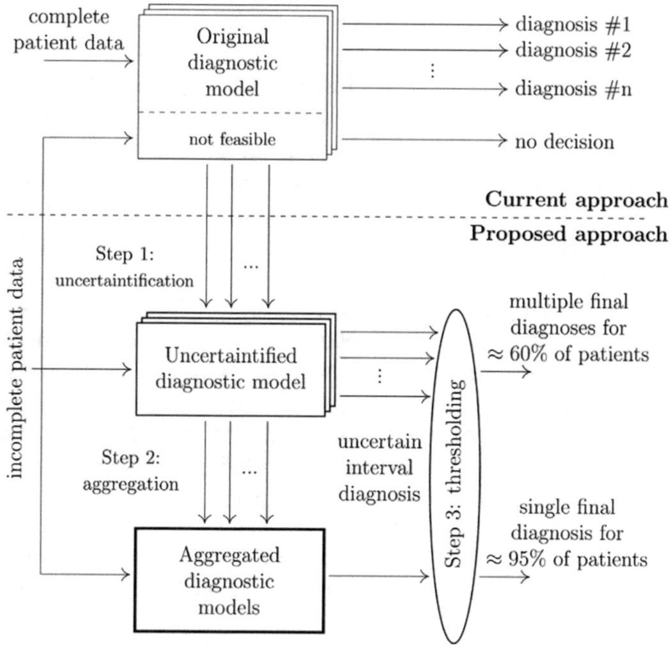

Fig. 6.1 OEA module in OvaExpert (cf. [8])

the best representatives of each class. Moreover, diverse methods of creating n-argument versions of binary aggregation operators were considered. If it comes to the class of possible and necessary aggregation functions we had to take into account the properties of comparability relations involved in the notions of these aggregation operators. Since, $\preceq_\pi$ is complete but it is not antisymmetric, we also took into account the width of intervals and the position of endpoints while creating the n-argument versions of these aggregation operators. The inputs were sorted increasingly and decreasingly. Since relation $\preceq_\nu$ is not complete, we also had to perform analogous procedure as for creating pos-aggregation functions. For both types of aggregation functions we considered methods of sorting involving $\preceq_\pi$ and $\preceq_\nu$. In the class of aggregation functions with respect to linear orders we have considered IVOWA operator (cf. Definition 2.23) with Xu and Yager order and we compared the obtained results to the ones for OEA from [8].

Concerning the classifier we used the same assumptions as in the original approach. To select the best aggregation operator from these returned in the training and testing phases, it was required that the following conditions should be satisfied:

- sensitivity $\geqslant 90\%$
- specificity $\geqslant 80\%$
- sensitivity > specificity
- decisiveness $< 100\%$.

The first two rules choose aggregation operators with high sensitivity and specificity values. The third rule reflects the fact that in a medical context sensitivity is more important than specificity. Since these two measures are correlated there may be some models (aggregation operators) that trade off sensitivity for specificity. Such models were rejected. Finally, models with 100% decisiveness were excluded in order not to impose diagnoses that lack sufficient justification. No decision, leading to further examinations, is better than a wrong decision. For classifiers that operate on poor quality data, such as incomplete data, in some applications it is necessary to consider a situation in which the classifier has insufficient information to make a credible decision. This may happen in medical applications when insufficiently certain decision can have serious consequences for the patient. In this case, it is better for the decision system to report inability to make a decision and suggest additional diagnostic tests. In the case of such classifiers, additional classes of errors $N0$ and $N1$ are introduced, describing the number of non-decisions in the positive and negative case, respectively. Decisiveness (*Dec* for short) determines in how many cases the classifier was able to make a decision, namely

$$Dec = \frac{TP + TN + FP + FN}{TP + TN + FP + FN + N0 + N1}.$$

It is worth to mention that in the case of classifiers that permit the option of not making a decision, sensitivity or specificity are calculated only for those cases in which the classifier has generated a decision. In this case all calculated measures for a given classifier should always be evaluated with reference to the decisiveness measure.

In OvaExpert interval representation of data was applied. This approach enabled effective decision-making, in spite of missing data. The novel approach applied in OvaExpert was to describe the value of each attribute of a patient by an interval, regardless of whether or not the description of the attribute was given. If the value was not provided, then the proposed representation had the form of a set containing all possible values for the attribute. If the value was given, it was represented by an interval reduced to a point. The main advantage of this approach is that all patients can be described in the same, uniform way and can be processed with the same diagnostic model. There are many different diagnostic models for ovarian tumor, which use different attributes describing the patient, and are therefore subject to different levels of uncertainty. The main idea was to improve the final diagnosis by taking advantage of the models diversity. For given n models $m_1, m_2, \ldots, m_n$, an aggregation operator is used and as result a new diagnosis is obtained that gathers information from the input models. As it was already mentioned, there are two possible modes of such aggregation. The first, called numerical, uses a single value that represents the whole interval (the most common choices are the interval's center, lower bound and upper bound). The interval mode utilizes the whole of the interval information. At the end some thresholding strategies are applied. They have the aim of converting a numerical or interval decision into a final diagnosis. For the numerical case there was only one class of thresholding strategies, i.e. thresholding with margin

Table 6.1 Cost matrix

	Predicted		
Actual	Benign	Malignant	NA
Benign	0	2.5	1
Malignant	5	0	2

$\varepsilon \in [-0.5, 0.5]$ given by

$$\tau_{\varepsilon}(a) = \begin{cases} B, & a < 0.5 - \varepsilon \\ M, & a \geqslant 0.5 + \varepsilon, \\ NA, & otherwise \end{cases} \quad a \in [0, 1].$$

For interval mode three thresholding strategies were evaluated. The first approach was to apply a numerical threshold to the interval representative. The second was to calculate the common part between intervals (cf. [10]). Finally, the third one was the interval version of thresholding with a margin given for each $\varepsilon \in [-0.5, 0.5]$ by

$$\tau_{\varepsilon}([a, b]) = \begin{cases} B, & b < 0.5 + \varepsilon \\ M, & a \geqslant 0.5 - \varepsilon, \\ NA, & otherwise \end{cases} \quad [a, b] \in L^I.$$

The notation B was used for benign tumor, M for the malignant one, and NA for the case, where there was no diagnosis.

In the medical diagnosis of ovarian tumors the situation when the system diagnoses a tumor as benign and, in fact, it was malignant causes much more significant effects for the patient, as opposed to the situation when the benign tumor is diagnosed as cancer. In such models the concept of cost matrix (cost function) is used where for each error type a weight (penalty) is assigned for a wrong decision. The quality value is the sum of costs (penalties) assigned to the classifier for making wrong decisions. If it comes to the cost matrix the costs have been selected in cooperation with experts in ovarian cancer diagnosis.

Table 6.1 presents costs (penalties) attributed to classifiers for incorrect decisions. Correct decisions did not receive a penalty. A classifier receives top penalty in the case if a patient with malign tumor is classified as a benign case. Penalty for the case if a patient with benign tumor was classified as malignant was half of it, as unjustified operation is still dangerous for a patient but death risk is much lower. Additionally, penalties for the classifier for failure to make a decision (NA) were used. The penalty is lower, as in such case the patient needs additional diagnostics and will probably be directed to a more experienced specialist who would make a correct diagnosis. However, penalties for lack of decision in positive (malignant) was twice as high as in the negative (benign) case. For more details we refer the reader to [8, 10].

6.2 Performance of Pos- and Nec-Aggregation Functions in OEA

In [19] some of the results connected with the performance of recently introduced classes of aggregation operators in medical decision support system OvaExpert (module OEA, cf. [8]) were given. The obtained results are promising and show usefulness of new classes in systems similar to OvaExpert. The new aggregation methods are connected with recently introduced possible and necessary aggregation functions [20] and also aggregation functions with respect to admissible linear orders [21]. These concepts of aggregation functions follow from diverse ways concerning comparability of intervals. Namely, these are possible and necessary comparability relations connected with epistemic and ontic settings of interval-valued calculus [22, 23] and the concept of linear orders introduced in [24]. It turned out that in OEA dataset the considered examples of aggregation operators are statistically comparable with the previously applied operators and in is some cases they obtained lower cost of prediction. There were applied source codes of OvaExpert available at GitHub [25]. However, the presented results were obtained with less number of repetitions in the training phase what may result in lower stability of the obtained results.

Here there are presented possible and necessary aggregation functions applied in OEA for which we obtained the best results. We have two pos-aggregation functions which are denoted by $\mathscr{A}_{pi1}$, $\mathscr{A}_{pi2}$, and the nec-aggregation function denoted by $\mathscr{A}_{nu}$. Let $\mathbf{x} = [\underline{x}, \overline{x}]$, $\mathbf{y} = [\underline{y}, \overline{y}]$. The following functions are interval-valued aggregation functions (non-representable) and they are also pos-aggregation functions (but they are not nec-aggregation functions), where

$$\mathscr{A}_{pi1}(\mathbf{x}, \mathbf{y}) = \begin{cases} [1, 1], & (\mathbf{x}, \mathbf{y}) = ([1, 1], [1, 1]) \\ \left[\frac{\underline{y} \frac{\underline{x}+\overline{x}}{2}}{2}, \frac{\overline{x}+\overline{y}}{2}\right], & \text{otherwise} \end{cases}$$

$$\mathscr{A}_{pi2}(\mathbf{x}, \mathbf{y}) = \begin{cases} [1, 1], & (\mathbf{x}, \mathbf{y}) = ([1, 1], [1, 1]) \\ \left[\frac{\underline{x} \frac{\underline{y}+\overline{y}}{2}}{2}, \frac{\overline{x}+\overline{y}}{2}\right], & \text{otherwise} \end{cases}$$

The following function is a nec-aggregation function (it is a pseudomax A_1A_2-representable aggregation function) but it is not a pos-aggregation function, where

$$\mathscr{A}_{nu}(\mathbf{x}, \mathbf{y}) = \left[\frac{\underline{x} + \underline{y}}{2}, \max\left(\frac{\underline{x} + \overline{y}}{2}, \frac{\overline{x} + \underline{y}}{2}\right)\right].$$

It was also considered IVOWA with the admissible linear order $\leq_{XY}$.

In Table 6.2 the measures of performance and cost matrix for the mentioned above the best representatives of the families of aggregation operators are presented. We analyze the results regarding accuracy, sensitivity, specificity and decisiveness. Accuracy (*Acc* for short, cf. Chap. 5) is a measure of the ratio of the number of correctly

Table 6.2 Comparison of performance for diverse aggregation operators in the test phase

Method	Cost matrix	Accuracy	Decisiveness	Sensitivity	Specificity
orig.Alc	186.5	0.889	0.206	0.941	0.842
orig.LR1	182.0	0.771	0.274	0.963	0.524
orig.LR2	161.5	0.828	0.331	0.971	0.609
orig.RMI	161.5	0.818	0.566	0.759	0.843
orig.SM	140.0	0.791	0.629	0.973	0.699
orig.Tim	161.5	0.904	0.474	0.667	0.956
unc.Alc	146.5	0.851	0.537	0.897	0.831
unc.LR1	152.0	0.721	0.594	0.978	0.517
unc.LR2	154.5	0.712	0.594	0.978	0.500
unc.RMI	128.0	0.858	0.766	0.767	0.885
unc.SM	111.5	0.790	0.926	0.935	0.733
unc.Tim	129.5	0.893	0.691	0.724	0.946
A_{pi}	74.0	0.851	0.960	0.980	0.797
A_{nu}	66.5	0.871	0.977	0.942	0.840
OEA	72.0	0.876	0.966	0.902	0.864
FSC	67.0	0.894	0.937	0.900	0.902
$IVOWA_{\leq_{XY}}$	65.5	0.887	0.960	0.922	0.872

classified objects to all evaluated objects (i.e. the probability of a proper classification). This is the most intuitive measure, but despite its simplicity it is not always the best measure. Especially in the case of unbalanced number of positive and negative cases in test sample. Additionally, accuracy does not work in situations where it is more important for us to have no errors of a given type. While, sensitivity specifies how many of the positive class objects are properly categorized. It can be interpreted as the probability that the classification will be correct, provided that the case is positive, i.e. the probability that the test performed for a cancer patient will show that the tumor is malignant. Specificity shows how often the model correctly classifies objects from the negative class. In other words, it is the probability that the classification will be correct, provided that the case was negative, i.e. the probability that a person with a benign tumor will show that the tumor is benign. Moreover, as it was already stated, decisiveness measure determines in how many cases the classifier was able to make a decision.

Table 6.2 presents the results for the original models [11–15] (denoted *orig.* for short) and the uncertaintified ones (denoted *unc.* for short) obtained in the test phase. The uncertaintified models are the original models adjusted to uncertainty data, i.e. data presented with the use of intervals. We also recall the results for the operator with the best results in OEA. It is an OWA operator with the central element of interval as a representative selector and $\tau_{0.025}$ as a threshold. We also give the results for FSC module (involving methods based on counting) which was presented in detail in [3]. Considered here aggregation operators (not included in the original system) were tested with diverse comparability relations if needed (for creating n-argument versions of operators or for ordering inputs). In Table 6.2 there are given results for the best obtained options. Since $\mathscr{A}_{pi1}$ and $\mathscr{A}_{pi2}$ gave similar results, we put only the data for $\mathscr{A}_{pi1}$ and denote it by $\mathscr{A}_{pi}$ which were obtained in interval mode with $\tau_{0.025}$, sorting with respect to $\preceq_{\pi}$ and width of intervals, the recurrence for obtaining n-argument versions of operators beginning from the right end. We see that these operators have weaker results concerning specificity than assumed in the system. For $\mathscr{A}_{nu}$ the results were obtained in interval mode with $\tau_{0.025}$, sorting with respect to $\preceq_{\pi}$ and width of intervals, the recurrence for obtaining n-argument versions of operators beginning from the right end (what is interesting, we obtained worse results for sorting with respect to the relation $\preceq_{\nu}$). Moreover, in this case we obtained better results for the inputs taken in the reverse form (i.e., $1 - x$) and the final result was again reversed. The results for IVOWA operator with Xu and Yager order (denoted in Table 6.2 by $IVOWA_{\leq_{XY}}$) were equally good both for the numerical and the interval mode. The threshold $\tau_{0.025}$ was used in both cases and in the numerical case the central element of interval was taken as a representative selector.

The McNemars test with Benjamini–Hochberg correction [8] was used to test new aggregation strategies with relation to the uncertaintified models. Detailed performance measures with 95% confidence intervals (CI) on test set for selected aggregation methods are presented in Table 6.3.

We may conclude that the presented representatives of newly tested aggregation operators in OEA obtained very good results in comparison to the operators and methods of ordering for inputs from [8]. Furthermore, in our experiment IVOWA operator with the Xu and Yager order applied for inputs ordering, obtained the lowest cost matrix among all the considered methods. However, the results were obtained for less number of repetitions than in the experiments presented in [8].

At the end of this chapter we are comparing the methods of interval modeling in the case of missing values applied in OEA (module of OvaExpert) and the methods pre-

Table 6.3 Detailed performance measures with 95% confidence intervals on test set for selected aggregation methods

Measure	$IVOWA_{\leq_{XY}}$	A_{pi}	A_{nu}
Accuracy	0.887	0.851	0.871
Accuracy 95% CI	(0.836–0.935)	(0.797–0.900)	(0.822–0.917)
Cost matrix	65.5	74.0	66.5
Cost matrix 95% CI	(40.000–94.733)	(50.000–98.733)	(44.500–93.233)
Decisiveness	0.960	0.960	0.977
Decisiveness 95% CI	(0.929–0.989)	(0.931–0.983)	(0.954–0.997)
Sensitivity	0.922	0.980	0.942
Sensitivity 95% CI	(0.837–0.981)	(0.939–1.000)	(0.865–1.000)
Specificity	0.872	0.797	0.840
Specificity 95% CI	(0.809–0.932)	(0.726–0.864)	(0.771–0.900)

sented in Chap. 4. In both classification algorithms interval modeling was performed at different stages. In OEA it was one of the first steps, namely uncertaintification of the prediction models, SM, Alc, LR1, LR2, Tim and RMI, expressed by real-valued functions, originally considered for ovarian tumor diagnosis [8]. Uncertaintification means enabling the diagnostic models to work with the interval representation of the patient data. A classical method of extending real functions to interval values [26] to obtain a new (uncertaintified) diagnostic model was used. The resultant interval represented all of the possible diagnoses that can be made based on a patient description in which every missing value has been replaced with all possible values for that attribute. In the approach presented in Chap. 4 interval modeling was applied at one of the final stages of algorithm, namely the classification by a single *k-NN* classifier, where on the base of object u random selection of r objects from training data and obtaining objects $u_1, \ldots, u_r$ was performed. Next classification of objects $u_1, \ldots, u_r$ and obtaining certainty coefficients $p_1, \ldots, p_r$ was done to determine intervals $[\min p_j, \max p_j]$, $j = 1, \ldots, r$ (cf. Algorithm 4). If it comes to the similarities between these two approaches, a fuzzy value p was transformed into interval-valued fuzzy value $[p, p]$ in the case of no missing values appearance in the conditional attributes (in Chap. 4) and in OEA at the stage of uncertaintification of the original diagnostic models.

To sum up, in this chapter we showed that application of aggregation operators defined with respect to the relations $\preceq_\pi$ and $\preceq_\nu$, as well as admissible linear orders, may result in very good results of cost prediction and other measures of the classifier quality in medical support systems applying aggregation methods.

References

1. OvaExpert project homepage: http://ovaexpert.pl/en/
2. Dyczkowski, K., Wójtowicz, A., Żywica, P., Stachowiak, A., Moszyński, R., Szubert, S.: An intelligent system for computer-aided ovarian tumor diagnosis. In: Intelligent Systems 2014, pp. 335–344. Springer International Publishing, Berlin (2015)
3. Dyczkowski, K.: Studies in Computational Intelligence. Intelligent medical decision support system based on imperfect information: the case of ovarian tumor diagnosis. Springer, Berlin (2018)
4. Moszyński, R., Żywica, P., Wójtowicz, A., Szubert, S., Sajdak, S., Stachowiak, A., Dyczkowski, K., Wygralak, M., Szpurek, D.: Menopausal status strongly influences the utility of predictive models in differential diagnosis of ovarian tumors: an external validation of selected diagnostic tools. Ginekol. Pol. **85**(12), 892–899 (2014)
5. Stachowiak, A., Dyczkowski, K., Wójtowicz, A., Żywica, P., Wygralak, M.: A bipolar view on medical diagnosis in ovaexpert system. In: Andreasen, T., Christiansen, H., Kacprzyk, J., et al. (eds.) Flexible Query Answering Systems 2015: Proceeding of the FQAS 2015, Cracow, Poland, October 26–28, 2015. Advances in Intelligent Systems and Computing, vol. 400, pp. 483–492. Springer International Publishing, Cham (2016)
6. Szubert, S., Wójtowicz, A., Moszyński, R., Żywica, P., Dyczkowski, K., Stachowiak, A., Sajdak, S., Szpurek, D., Alcázar, J.L.: External validation of the IOTA ADNEX model performed by two independent gynecologic centers. Gynecol. Oncol. **142**(3), 490–495 (2016)
7. Wójtowicz, A., Żywica, P., Szarzyński, K., Moszyński, R., Szubert, S., Dyczkowski, K., Stachowiak, A., Szpurek, D., Wygralak, M.: Dealing with uncertinity in ovarian tumor diagnosis.

In: Modern Approaches in Fuzzy Sets, Intuitionistic Fuzzy Sets, Generalized Nets and Related Topics. Vol. II: Applications, pp. 151–158. SRI PAS, Warszawa (2014)
8. Wójtowicz, A., Żywica, P., Stachowiak, A., Dyczkowski, K.: Solving the problem of incomplete data in medical diagnosis via interval modeling. Appl. Soft Comput. **47**, 424–437 (2016)
9. Żywica, P., Wójtowicz, A., Stachowiak, A., Dyczkowski, K.: Improving medical decisions under incomplete data using intervalvalued fuzzy aggregation. In: Proceeding of the IFSA-EUSFLAT 2015, pp. 577–584. Atlantis Press, Asturias (2015)
10. Żywica, P., Dyczkowski, K., Wójtowicz, A., Stachowiak, A., Szubert, S., Moszyński, R.: Development of a fuzzy-driven system for ovarian tumor diagnosis. Biocybern. Biomed. Eng. **36**(4), 632–643 (2016)
11. Alcázar, J.L., Mercé, L.T., et al.: A new scoring system to differentiate benign from malignant adnexal masses. Obstet. Gynecol. Surv. **58**(7), 462–463 (2003)
12. Szpurek, D., Moszyński, R., et al.: An ultrasonographic morphological index forprediction of ovarian tumor malignancy. Eur. J. Gynaecol. Oncol. **26**(1), 51–54 (2005)
13. Jacobs, I., Oram, D., et al.: A risk of malignancy index incorporating CA 125, ultrasound and menopausal status for the accurate preoperative diagnosis ofovarian cancer. BJOG **97**(10), 922–929 (1990)
14. Timmerman, D., Bourne, T.H., et al.: A comparison of methods for preoperative discrimination between malignant and benign adnexal masses: the development of a new logistic regression model. Am. J. Obstet. Gynecol. **181**(1), 57–65 (1999)
15. Timmerman, D., Testa, A.C., et al.: Logistic regression model to distinguish between the benign and malignant adnexal mass before surgery: amulticenter study by the international ovarian tumor analysis group. J. Clin. Oncol. **23**(34), 8794–8801 (2005)
16. Bentkowska, U., Król, A.: Preservation of fuzzy relation properties based on fuzzy conjunctions and disjunctions during aggregation process. Fuzzy Sets Syst. **291**, 98–113 (2016)
17. Bentkowska, U.: Aggregation of diverse types of fuzzy orders for decision making problems. Inf. Sci. **424**, 317–336 (2018)
18. De Miguel, L., Bustince, H., Pękala, B., Bentkowska, U., Da Silva, I., Bedregal, B., Mesiar, R., Ochoa, G.: Interval-valued Atanassov intuitionistic OWA aggregations using admissible linear orders and their application to decision making. IEEE Trans. Fuzzy Syst. **24**(6), 1586–1597 (2016)
19. Bentkowska, U., Pękala, B.: Diverse classes of interval-valued aggregation functions in medical diagnosis support. In: Medina, J. et al. (eds.) IPMU 2018, pp. 391–403. Springer International Publishing AG, part of Springer, CCIS 855 (2018)
20. Bentkowska, U.: New types of aggregation functions for interval-valued fuzzy setting and preservation of pos-B and nec-B-transitivity in decision making problems. Inf. Sci. **424**, 385–399 (2018)
21. Zapata, H., Bustince, H., Montes, S., Bedregal, B., Dimuro, G.P., Takáč, Z., Baczyński, M., Fernandez, J.: Interval-valued implications and interval-valued strong equality index with admissible orders. Int. J. Approx. Reason. **88**, 91–109 (2017)
22. Dubois, D., Prade, H.: Possibility Theory. Plenum Press, New York (1988)
23. Dubois, D., Prade, H.: Gradualness, uncertainty and bipolarity: making sense of fuzzy sets. Fuzzy Sets Syst. **192**, 3–24 (2012)
24. Bustince, H., Fernandez, J., Kolesárová, A., Mesiar, R.: Generation of linear orders for intervals by means of aggregation functions. Fuzzy Sets Syst. **220**, 69–77 (2013)
25. https://github.com/ovaexpert
26. Moore, R.E.: Interval Analysis, vol. 4. Prentice-Hall, Englewood Cliffs (1966)

Chapter 7
Summary

I love the fact that I get something new to do almost every day and have new challenges.

Jim Lee

This monograph presents theory concerning interval-valued fuzzy calculus, especially aggregation operators among which there are placed recently introduced pos-aggregation functions and nec-aggregation functions. Moreover, applications of interval-valued fuzzy methods (and more generally interval modeling), especially mentioned so far interval-valued aggregation functions in classification are provided. The presented algorithms may support decision processes for example in medical diagnosis. It was shown that applying interval-valued methods make it possible to improve classification results in situation when there is a large number of missing values or there is a large number of attributes in the considered data sets. The mentioned methods were applied in a few experiments, involving machine learning methods, whose results are presenin this monograph. The obtained results of the described algorithms confirm their usefulness. They allow effective classification in the case of uncertainty (imprecise or incomplete information). The source codes of the presented classification algorithms (from Chaps. 4 and 5) are available at [1].

It is worth mentioning here the ongoing research concerning the considered algorithms. Namely, we would like to do research with multi-valued classifiers, in the k-NN algorithm apply other distances than the Euclidean one and use in the experiments more sophisticated examples of interval-valued aggregation operators. As we see there are many future directions of development for the presented here research. Moreover, we would like to study what are the theoretical properties of the applied in the experiments aggregation operators (mainly pos-aggregation functions and nec-aggregation functions, cf. [2]) that allowed to obtain better results of classification.

Another interesting problem concerning classification is the choice of classification methods depending on a given data set. For example, in [3] were presented automatic extraction method to determine the domains of competence of a classifier

U. Bentkowska, *Interval-Valued Methods in Classifications and Decisions*,
Studies in Fuzziness and Soft Computing 378,
https://doi.org/10.1007/978-3-030-12927-9_7

using a set of data complexity measures proposed for the task of classification. In that paper as a representative for classifiers to analyze the proposal, three classical but different algorithms were used: C4.5, SVM and k-NN. Similarly, analogous question arises in the case when in classification the aggregation methods are involved, i.e. the question which aggregation operator is the most suitable for a given set of data. There were some preliminary approaches to consider this matter presented in [4] for aggregation functions acting on [0, 1]. In that paper were performed studies whether there are characteristics of the data sets (its properties or measures) that allow one to know whether an aggregation function will work better then others or not. This research was focused on diverse classes of Choquet integral, which belong to the family of aggregation functions, where on the considered data sets some domains of competence of the approaches were extracted (the method proposed in [3] was taken into account). From our point of view, for the future research, interval-valued aggregation functions would be of interest, especially pos-aggregation functions and nec-aggregation functions recently introduced and being in the center of attention in this monograph.

Moreover, in [5] were presented a range of regression techniques specifically tailored to building aggregation operators from empirical data. These techniques identify optimal parameters of aggregation operators from various classes (triangular norms, uninorms, copulas, ordered weighted aggregation OWA, generalized means, compensatory and general aggregation operators), while allowing one to preserve specific properties such as commutativity or associativity. One of the latest papers devoted to this topic is [6], where also the problem of missing data was taken into account as the one common in real-world applications of supervised machine learning such as classification and regression. There were proposed optimization methods for learning the weights of quasi-arithmetic means in the context of data with missing values. Other contributions devoted to the topic of fitting aggregation functions to data, with the focus on weighted arithmetic means, are [7, 8]. In these papers diverse methods, such as linearization, regularization and idempotization, were compared. This is another interesting topic that could be considered for pos- and nec-aggregation functions. Methods for construction of aggregation functions from data by minimizing the least absolute deviation criterion were presented in [9]. Various instances of such problems as linear programming ones were formulated. What is interesting, the cases in which the data were provided as intervals were considered. However, the aggregation functions on [0, 1] were constructed. While fitting aggregation functions to data, preservation of outputs ordering is also important, perhaps even more than fitting the output values. The reason quoted in [10] is that humans are better at ranking the alternatives than expressing their preferences as numbers. Thus preservation of outputs ordering may be equally or more important than fitting these outputs. This problem was addressed in [9, 11]. For the future research it would be interesting to find methods of fitting interval-valued aggregation operators to interval data and also to build such aggregation operators that are able to preserve ordering of the interval data.

References

1. http://diagres.ur.edu.pl/~fuzzydataminer/
2. Bentkowska, U.: New types of aggregation functions for interval-valued fuzzy setting and preservation of pos-B and nec-B-transitivity in decision making problems. Inf. Sci. **424**, 385–399 (2018)
3. Luengo, J., Herrera, F.: An automatic extraction method of the domains of competence for learning classifiers using data complexity measures. Knowl. Inf. Syst. **42**, 147–180 (2015)
4. Lucca, G., Sanz, J.A., Dimuro, G.P., Bedregal, B., Bustince, H.: Analyzing the behavior of aggregation and pre-aggregation functions in fuzzy rule-based classification systems with data complexity measures. In: Kacprzyk, J., et al. (eds.) Advances in Intelligent Systems and Computing, vol. 642, pp. 443–455. Springer AG, Cham (2018)
5. Beliakov, G.: How to build aggregation operators from data. Int. J. Intell. Syst. **18**, 903–923 (2003)
6. Beliakov, G., Gomez, D., James, J., Montero, J., Rodriguez, J.T.: Approaches to learning strictly-stable weights for data with missing values. Fuzzy Sets Syst. **325**, 97–113 (2017)
7. Bartoszuk, M., Beliakov, G., Gagolewski, M., James, S.: Fitting aggregation functions to data: part I linearization and regularization. In: Carvalho, J.P., et al. (eds.) Information Processing and Management of Uncertainty in Knowledge-Based Systems, Part II, CCIS 611, IPMU 2016, Eindhoven, Netherlands, 2016, pp. 767–779. Springer International Publishing, Switzerland (2016)
8. Bartoszuk, M., Beliakov, G., Gagolewski, M., James, S.: Fitting aggregation functions to data: part II idempotization. In: Carvalho, J.P., et al. (eds.) Information Processing and Management of Uncertainty in Knowledge-Based Systems, Part II, CCIS 611, IPMU 2016, Eindhoven, Netherlands, 2016, pp. 780–789. Springer International Publishing, Switzerland (2016)
9. Beliakov, G.: Construction of aggregation functions from data using linear programming. Fuzzy Sets Syst. **160**, 65–75 (2009)
10. Kaymak, U. Van Nauta Lemke, H.R.: Selecting an aggregation operator for fuzzy decision making. In: Proceedings of the 1994 IEEE 3rd International Fuzzy Systems Conference, vol. 2, pp. 1418–1422. Orlando, FL (1994)
11. Beliakov, G., Calvo, T.: Constructions of aggregation operators that preserve ordering of the data. In: Štěpnička, M., Novák, V., Bodenhofer, U. (eds.) New Dimensions in Fuzzy Logic and Related Technologies. Proceedings of the 5th EUSFLAT Conference, Ostrava, Czech Republic, 11–14 Sept 2007, pp. 61–66. Universitas Ostraviensis, Ostrava (2007)

Chapter 8
Tables with the Results of Experiments

There is strength in numbers, but organizing those numbers is one of the great challenges.

John C. Mather

In this chapter the numerical values obtained as the results of experiments are presented. They are organized in tables consisting of appropriate data depending on the type of experiment. The provided tables refer to experiments described in Chaps. 4 and 5.

The data referring to experiments described in Chap. 4 (cf. Tables 8.1, 8.2 and 8.3) present the type of algorithm (with adequate parameters such as aggregation operator for the method F and the value of parameter k for the k-NN algorithm in the method C) as well as the level of missing values and average AUC with the standard deviation obtained in the successive repetitions.

Similarly, the results from Chap. 5 are gathered in Tables 8.4, 8.5 and 8.6, where the value NoClassHoriz corresponds to the value h and the value NoClassVert corresponds to the value v (cf. Sect. 5.3). Moreover, there is given information about the applied method (for the method F there is added the symbol of the aggregation operator applied, cf. (4.1)–(4.6)), the value k for the k-NN classifier, and average values accompanied with the standard deviation of accuracy ACC, sensitivity SN, specificity SP (for 20 repetitions of experiments).

U. Bentkowska, *Interval-Valued Methods in Classifications and Decisions*,
Studies in Fuzziness and Soft Computing 378,
https://doi.org/10.1007/978-3-030-12927-9_8

Table 8.1 The average AUC for classification algorithms, part I

Missing	Banknote		Biodeg		Breast cancer		Diabetes	
	Method	AUC	Method	AUC	Method	AUC	Method	AUC
0.0	C, k = 15	1.0 ± 0.001	F, $\mathscr{A}_6$	0.903 ± 0.014	C, k = 10	0.991 ± 0.006	M	0.805 ± 0.013
0.0	C, k = 20	1.0 ± 0.0	F, $\mathscr{A}_1$	0.902 ± 0.007	C, k = 15	0.991 ± 0.004	C, k = 30	0.804 ± 0.018
0.0	C, k = 30	1.0 ± 0.0	F, $\mathscr{A}_2$	0.902 ± 0.01	M	0.991 ± 0.005	F, $\mathscr{A}_6$	0.804 ± 0.015
0.0	M	1.0 ± 0.0	F, $\mathscr{A}_3$	0.902 ± 0.014	F, $\mathscr{A}_1$	0.991 ± 0.004	F, $\mathscr{A}_1$	0.8 ± 0.011
0.0	F, $\mathscr{A}_1$	1.0 ± 0.0	F, $\mathscr{A}_5$	0.901 ± 0.012	F, $\mathscr{A}_3$	0.99 ± 0.003	F, $\mathscr{A}_2$	0.8 ± 0.02
0.0	F, $\mathscr{A}_2$	1.0 ± 0.0	M	0.899 ± 0.008	C, k = 5	0.989 ± 0.006	F, $\mathscr{A}_5$	0.8 ± 0.018
0.0	F, $\mathscr{A}_3$	1.0 ± 0.0	F, $\mathscr{A}_4$	0.895 ± 0.011	C, k = 20	0.989 ± 0.005	F, $\mathscr{A}_3$	0.798 ± 0.011
0.0	F, $\mathscr{A}_4$	1.0 ± 0.0	C, k = 10	0.891 ± 0.009	C, k = 3	0.988 ± 0.004	F, $\mathscr{A}_4$	0.796 ± 0.017
0.0	F, $\mathscr{A}_5$	1.0 ± 0.0	C, k = 15	0.891 ± 0.022	F, $\mathscr{A}_2$	0.988 ± 0.008	C, k = 20	0.795 ± 0.017
0.0	F, $\mathscr{A}_6$	1.0 ± 0.0	C, k = 30	0.89 ± 0.009	F, $\mathscr{A}_5$	0.988 ± 0.005	C, k = 15	0.793 ± 0.021
0.0	C, k = 3	0.999 ± 0.001	C, k = 20	0.887 ± 0.01	F, $\mathscr{A}_6$	0.987 ± 0.006	C, k = 10	0.785 ± 0.009
0.0	C, k = 5	0.999 ± 0.001	C, k = 5	0.886 ± 0.013	C, k = 30	0.986 ± 0.009	C, k = 5	0.762 ± 0.024
0.0	C, k = 10	0.999 ± 0.001	C, k = 3	0.87 ± 0.015	F, $\mathscr{A}_4$	0.986 ± 0.008	C, k = 3	0.744 ± 0.019
0.0	C, k = 1	0.998 ± 0.001	C, k = 1	0.814 ± 0.013	C, k = 1	0.946 ± 0.013	C, k = 1	0.663 ± 0.018
0.05	F, $\mathscr{A}_1$	0.974 ± 0.015	F, $\mathscr{A}_4$	0.818 ± 0.049	F, $\mathscr{A}_2$	0.987 ± 0.005	F, $\mathscr{A}_1$	0.762 ± 0.013
0.05	F, $\mathscr{A}_3$	0.971 ± 0.02	F, $\mathscr{A}_1$	0.806 ± 0.058	C, k = 15	0.985 ± 0.004	F, $\mathscr{A}_4$	0.752 ± 0.032
0.05	F, $\mathscr{A}_4$	0.971 ± 0.022	F, $\mathscr{A}_2$	0.79 ± 0.054	F, $\mathscr{A}_1$	0.985 ± 0.009	F, $\mathscr{A}_3$	0.749 ± 0.029
0.05	F, $\mathscr{A}_6$	0.97 ± 0.02	F, $\mathscr{A}_3$	0.79 ± 0.056	F, $\mathscr{A}_4$	0.985 ± 0.006	F, $\mathscr{A}_6$	0.748 ± 0.036
0.05	F, $\mathscr{A}_2$	0.969 ± 0.016	F, $\mathscr{A}_6$	0.785 ± 0.066	F, $\mathscr{A}_6$	0.985 ± 0.006	F, $\mathscr{A}_2$	0.746 ± 0.054
0.05	C, k = 15	0.933 ± 0.037	F, $\mathscr{A}_5$	0.766 ± 0.035	M	0.984 ± 0.007	C, k = 20	0.705 ± 0.031
0.05	C, k = 30	0.93 ± 0.044	M	0.738 ± 0.058	F, $\mathscr{A}_3$	0.984 ± 0.006	M	0.703 ± 0.042

(continued)

Table 8.1 (continued)

	Banknote		Biodeg		Breast cancer		Diabetes	
Missing	Method	AUC	Method	AUC	Method	AUC	Method	AUC
0.05	M	0.928 ± 0.041	C, k = 10	0.736 ± 0.035	C, k = 30	0.983 ± 0.007	C, k = 15	0.693 ± 0.04
0.05	C, k = 20	0.925 ± 0.044	C, k = 15	0.725 ± 0.048	C, k = 20	0.981 ± 0.005	C, k = 10	0.689 ± 0.047
0.05	F, $\mathscr{A}_5$	0.92 ± 0.031	C, k = 30	0.72 ± 0.07	C, k = 10	0.979 ± 0.008	C, k = 30	0.686 ± 0.045
0.05	C, k = 10	0.905 ± 0.042	C, k = 20	0.708 ± 0.059	F, $\mathscr{A}_5$	0.972 ± 0.01	F, $\mathscr{A}_5$	0.677 ± 0.038
0.05	C, k = 5	0.903 ± 0.042	C, k = 5	0.696 ± 0.055	C, k = 5	0.969 ± 0.016	C, k = 5	0.651 ± 0.039
0.05	C, k = 3	0.887 ± 0.051	C, k = 3	0.682 ± 0.052	C, k = 3	0.949 ± 0.022	C, k = 3	0.648 ± 0.042
0.05	C, k = 1	0.868 ± 0.051	C, k = 1	0.573 ± 0.055	C, k = 1	0.89 ± 0.03	C, k = 1	0.609 ± 0.035
0.1	F, $\mathscr{A}_6$	0.939 ± 0.037	F, $\mathscr{A}_3$	0.761 ± 0.058	F, $\mathscr{A}_2$	0.984 ± 0.005	F, $\mathscr{A}_2$	0.738 ± 0.052
0.1	F, $\mathscr{A}_3$	0.932 ± 0.043	F, $\mathscr{A}_4$	0.759 ± 0.06	F, $\mathscr{A}_1$	0.98 ± 0.01	F, $\mathscr{A}_3$	0.726 ± 0.04
0.1	F, $\mathscr{A}_1$	0.929 ± 0.041	F, $\mathscr{A}_2$	0.753 ± 0.079	F, $\mathscr{A}_3$	0.979 ± 0.007	F, $\mathscr{A}_1$	0.721 ± 0.039
0.1	F, $\mathscr{A}_4$	0.929 ± 0.039	F, $\mathscr{A}_1$	0.745 ± 0.065	F, $\mathscr{A}_6$	0.978 ± 0.009	F, $\mathscr{A}_6$	0.71 ± 0.054
0.1	F, $\mathscr{A}_2$	0.911 ± 0.048	F, $\mathscr{A}_6$	0.73 ± 0.083	C, k = 1	0.976 ± 0.013	F, $\mathscr{A}_4$	0.704 ± 0.053
0.1	C, k = 15	0.863 ± 0.064	C, k = 15	0.698 ±0.045	F, $\mathscr{A}_4$	0.974 ± 0.013	C, k = 30	0.657 ± 0.043
0.1	F, $\mathscr{A}_5$	0.853 ± 0.05	F, $\mathscr{A}_5$	0.692 ±0.043	C, k = 20	0.973 ± 0.014	F, $\mathscr{A}_5$	0.652 ± 0.055
0.1	C, k = 20	0.846 ± 0.062	C, k = 5	0.679 ± 0.055	M	0.972 ± 0.014	C, k = 10	0.643 ± 0.051
0.1	M	0.846 ± 0.076	M	0.674 ± 0.059	C, k = 15	0.97 ± 0.012	C, k = 20	0.638 ± 0.057
0.1	C, k = 30	0.832 ± 0.081	C, k = 20	0.667 ± 0.063	F, $\mathscr{A}_5$	0.958 ± 0.02	C, k = 15	0.628 ± 0.048
0.1	C, k = 5	0.823 ± 0.068	C, k = 30	0.658 ± 0.083	C, k = 10	0.957 ± 0.018	M	0.626 ± 0.043
0.1	C, k = 3	0.817 ± 0.078	C, k = 10	0.65 ± 0.08	C, k = 5	0.956 ± 0.026	C, k = 5	0.602 ± 0.076
0.1	C, k = 10	0.815 ± 0.089	C, k = 3	0.611 ± 0.068	C, k = 3	0.892 ± 0.047	C, k = 3	0.587 ± 0.047
0.1	C, k = 1	0.776 ± 0.089	C, k = 1	0.591 ± 0.033	C, k = 1	0.785 ± 0.089	C, k = 1	0.578 ± 0.03

(continued)

Table 8.1 (continued)

	Banknote		Biodeg		Breast cancer		Diabetes	
Missing	Method	AUC	Method	AUC	Method	AUC	Method	AUC
0.3	F, $\mathscr{A}_6$	0.756 ± 0.107	F, $\mathscr{A}_2$	0.644 ± 0.082	F, $\mathscr{A}_3$	0.91 ± 0.058	F, $\mathscr{A}_4$	0.635 ± 0.056
0.3	F, $\mathscr{A}_3$	0.754 ± 0.101	F, $\mathscr{A}_3$	0.629 ± 0.097	F, $\mathscr{A}_6$	0.903 ± 0.06	F, $\mathscr{A}_3$	0.625 ± 0.078
0.3	F, $\mathscr{A}_4$	0.752 ± 0.107	F, $\mathscr{A}_4$	0.624 ± 0.099	F, $\mathscr{A}_1$	0.901 ± 0.061	F, $\mathscr{A}_1$	0.62 ± 0.08
0.3	F, $\mathscr{A}_2$	0.748 ± 0.11	F, $\mathscr{A}_6$	0.623 ± 0.093	F, $\mathscr{A}_4$	0.892 ± 0.072	F, $\mathscr{A}_6$	0.62 ± 0.066
0.3	F, $\mathscr{A}_1$	0.74 ± 0.118	F, $\mathscr{A}_1$	0.622 ± 0.087	F, $\mathscr{A}_2$	0.891 ± 0.074	F, $\mathscr{A}_2$	0.615 ± 0.083
0.3	C, k = 30	0.681 ± 0.082	F, $\mathscr{A}_5$	0.595 ± 0.047	C, k = 30	0.879 ± 0.063	C, k = 15	0.606 ± 0.041
0.3	F, $\mathscr{A}_5$	0.681 ± 0.089	C, k = 30	0.591 ± 0.061	M	0.853 ± 0.072	F, $\mathscr{A}_5$	0.591 ± 0.05
0.3	C, k = 20	0.674 ± 0.065	C, k = 20	0.59 ± 0.073	C, k = 15	0.85 ± 0.066	C, k = 10	0.574 ± 0.065
0.3	C, k = 10	0.654 ± 0.083	C, k = 10	0.587 ± 0.063	C, k = 20	0.842 ± 0.078	C, k = 30	0.568 ± 0.052
0.3	C, k = 15	0.642 ± 0.091	M	0.58 ± 0.06	F, $\mathscr{A}_5$	0.832 ± 0.065	C, k = 3	0.559 ± 0.026
0.3	M	0.637 ± 0.104	C, k = 5	0.568 ± 0.047	C, k = 10	0.818 ± 0.075	C, k = 20	0.549 ± 0.044
0.3	C, k = 3	0.629 ± 0.071	C, k = 15	0.553 ± 0.061	C, k = 5	0.78 ± 0.07	M	0.545 ± 0.05
0.3	C, k = 5	0.614 ± 0.073	C, k = 3	0.526 ± 0.02	C, k = 3	0.711 ± 0.106	C, k = 5	0.526 ± 0.027
0.3	C, k = 1	0.587 ± 0.058	C, k = 1	0.511 ± 0.031	C, k = 1	0.629 ± 0.069	C, k = 1	0.525 ± 0.031
0.5	M	0.655 ± 0.152	F, $\mathscr{A}_4$	0.581 ± 0.064	F, $\mathscr{A}_6$	0.766 ± 0.128	F, $\mathscr{A}_1$	0.572 ± 0.072
0.5	F, $\mathscr{A}_6$	0.641 ± 0.11	M	0.577 ± 0.073	F, $\mathscr{A}_4$	0.764 ± 0.141	F, $\mathscr{A}_3$	0.568 ± 0.055
0.5	C, k = 30	0.64 ± 0.074	F, $\mathscr{A}_1$	0.575 ± 0.078	F, $\mathscr{A}_3$	0.761 ± 0.137	C, k = 15	0.561 ± 0.041
0.5	F, $\mathscr{A}_3$	0.634 ± 0.092	F, $\mathscr{A}_3$	0.565 ± 0.083	C, k = 30	0.753 ± 0.141	F, $\mathscr{A}_2$	0.56 ± 0.061
0.5	F, $\mathscr{A}_4$	0.631 ± 0.103	F, $\mathscr{A}_2$	0.561 ± 0.055	M	0.752 ± 0.103	F, $\mathscr{A}_4$	0.559 ± 0.066
0.5	F, $\mathscr{A}_1$	0.624 ± 0.098	F, $\mathscr{A}_6$	0.561 ± 0.058	F, $\mathscr{A}_1$	0.751 ± 0.135	F, $\mathscr{A}_6$	0.557 ± 0.056
0.5	F, $\mathscr{A}_2$	0.618 ± 0.102	C, k = 30	0.55 ± 0.066	F, $\mathscr{A}_2$	0.75 ± 0.141	M	0.555 ± 0.045

(continued)

Table 8.1 (continued)

	Banknote		Biodeg		Breast cancer		Diabetes	
Missing	Method	AUC	Method	AUC	Method	AUC	Method	AUC
0.5	C, k = 15	0.604 ± 0.083	C, k = 20	0.544 ± 0.043	C, k = 5	0.724 ± 0.095	F, $\mathscr{A}_5$	0.554 ± 0.052
0.5	C, k = 20	0.604 ± 0.085	C, k = 15	0.535 ± 0.053	C, k = 15	0.719 ± 0.092	C, k = 20	0.53 ± 0.029
0.5	C, k = 10	0.592 ± 0.051	F, $\mathscr{A}_5$	0.53 ± 0.038	F, $\mathscr{A}_5$	0.705 ± 0.085	C, k = 3	0.525 ± 0.033
0.5	F, $\mathscr{A}_5$	0.576 ± 0.084	C, k = 10	0.507 ± 0.078	C, k = 20	0.699 ± 0.12	C, k = 5	0.524 ± 0.037
0.5	C, k = 5	0.568 ± 0.068	C, k = 5	0.503 ± 0.026	C, k = 10	0.692 ± 0.085	C, k = 10	0.52 ± 0.042
0.5	C, k = 1	0.541 ± 0.04	C, k = 1	0.501 ± 0.006	C, k = 3	0.601 ± 0.095	C, k = 1	0.515 ± 0.02
0.5	C, k = 3	0.539 ± 0.038	C, k = 3	0.497 ± 0.024	C, k = 1	0.578 ± 0.061	C, k = 30	0.514 ± 0.04

Table 8.2 The average AUC for classification algorithms, part II

	German		Ozone		Parkinson	
Missing	Method	AUC	Method	AUC	Method	AUC
0.0	C, k = 30	0.752 ± 0.028	C, k = 15	0.82 ± 0.042	C, k = 30	1.0 ± 0.0
0.0	M	0.742 ± 0.01	F, $\mathscr{A}_4$	0.816 ± 0.036	M	1.0 ± 0.0
0.0	F, $\mathscr{A}_4$	0.736 ± 0.01	C, k = 20	0.811 ± 0.049	F, $\mathscr{A}_1$	1.0 ± 0.0
0.0	F, $\mathscr{A}_3$	0.735 ± 0.015	F, $\mathscr{A}_2$	0.809 ± 0.048	F, $\mathscr{A}_2$	1.0 ± 0.001
0.0	F, $\mathscr{A}_5$	0.735 ± 0.017	F, $\mathscr{A}_3$	0.805 ± 0.023	F, $\mathscr{A}_3$	1.0 ± 0.0
0.0	F, $\mathscr{A}_1$	0.734 ± 0.017	F, $\mathscr{A}_6$	0.805 ± 0.031	F, $\mathscr{A}_4$	1.0 ± 0.0
0.0	C, k = 15	0.733 ± 0.014	F, $\mathscr{A}_5$	0.788 ± 0.024	F, $\mathscr{A}_5$	1.0 ± 0.0
0.0	C, k = 20	0.733 ± 0.012	F, $\mathscr{A}_1$	0.785 ± 0.041	F, $\mathscr{A}_6$	1.0 ± 0.0
0.0	F, $\mathscr{A}_2$	0.731 ± 0.018	C, k = 30	0.784 ± 0.038	C, k = 5	0.999 ± 0.001
0.0	F, $\mathscr{A}_6$	0.728 ± 0.023	M	0.777 ± 0.034	C, k = 10	0.999 ± 0.001
0.0	C, k = 10	0.722 ± 0.016	C, k = 10	0.763 ± 0.034	C, k = 15	0.999 ± 0.0

(continued)

Table 8.2 (continued)

	German		Ozone		Parkinson	
Missing	Method	AUC	Method	AUC	Method	AUC
0.0	C, k = 5	0.69 ± 0.024	C, k = 5	0.692 ± 0.028	C, k = 20	0.999 ± 0.0
0.0	C, k = 3	0.676 ± 0.021	C, k = 3	0.662 ± 0.022	C, k = 3	0.997 ± 0.002
0.0	C, k = 1	0.619 ± 0.018	C, k = 1	0.579 ± 0.02	C, k = 1	0.985 ± 0.004
0.05	F, $\mathscr{A}_1$	0.729 ± 0.023	F, $\mathscr{A}_6$	0.755 ± 0.104	F, $\mathscr{A}_2$	0.98 ± 0.017
0.05	C, k = 20	0.712 ± 0.011	F, $\mathscr{A}_1$	0.75 ± 0.118	F, $\mathscr{A}_1$	0.979 ± 0.014
0.05	F, $\mathscr{A}_5$	0.712 ± 0.03	F, $\mathscr{A}_2$	0.735 ± 0.059	F, $\mathscr{A}_4$	0.974 ± 0.022
0.05	F, $\mathscr{A}_3$	0.708 ± 0.02	F, $\mathscr{A}_3$	0.715 ± 0.126	F, $\mathscr{A}_3$	0.972 ± 0.02
0.05	F, $\mathscr{A}_6$	0.704 ± 0.028	F, $\mathscr{A}_5$	0.698 ± 0.129	F, $\mathscr{A}_6$	0.972 ± 0.015
0.05	M	0.703 ± 0.025	F, $\mathscr{A}_4$	0.661 ± 0.124	C, k = 30	0.939 ± 0.031
0.05	F, $\mathscr{A}_4$	0.701 ± 0.047	C, k = 30	0.653 ± 0.092	M	0.934 ± 0.022
0.05	C, k = 30	0.699 ± 0.04	M	0.645 ± 0.049	C, k = 20	0.929 ± 0.04
0.05	C, k = 15	0.693 ± 0.028	C, k = 15	0.636 ± 0.075	C, k = 15	0.926 ± 0.039
0.05	F, $\mathscr{A}_2$	0.686 ± 0.039	C, k = 20	0.629 ± 0.046	F, $\mathscr{A}_5$	0.922 ± 0.021
0.05	C, k = 10	0.674 ± 0.038	C, k = 5	0.588 ± 0.067	C, k = 5	0.917 ± 0.033
0.05	C, k = 5	0.643 ± 0.029	C, k = 10	0.553 ± 0.07	C, k = 10	0.911 ± 0.036
0.05	C, k = 3	0.622 ± 0.026	C, k = 3	0.519 ± 0.032	C, k = 3	0.883 ± 0.044
0.05	C, k = 1	0.602 ± 0.031	C, k = 1	0.503 ± 0.006	C, k = 1	0.816 ± 0.083
0.1	F, $\mathscr{A}_1$	0.698 ± 0.04	F, $\mathscr{A}_5$	0.764 ± 0.096	F, $\mathscr{A}_4$	0.937 ± 0.042
0.1	F, $\mathscr{A}_4$	0.69 ± 0.033	F, $\mathscr{A}_3$	0.668 ± 0.109	F, $\mathscr{A}_2$	0.935 ± 0.049
0.1	F, $\mathscr{A}_6$	0.687 ± 0.034	F, $\mathscr{A}_4$	0.656 ± 0.113	F, $\mathscr{A}_1$	0.934 ± 0.04
0.1	F, $\mathscr{A}_3$	0.683 ± 0.044	C, k = 15	0.634 ± 0.087	F, $\mathscr{A}_3$	0.93 ± 0.046

(continued)

Table 8.2 (continued)

	German		Ozone		Parkinson	
Missing	Method	AUC	Method	AUC	Method	AUC
0.1	C, k = 30	0.669 ± 0.059	C, k = 20	0.626 ± 0.078	F, $\mathscr{A}_6$	0.927 ± 0.053
0.1	F, $\mathscr{A}_5$	0.669 ± 0.035	F, $\mathscr{A}_1$	0.62 ± 0.118	C, k = 15	0.877 ± 0.053
0.1	M	0.667 ± 0.045	F, $\mathscr{A}_5$	0.601 ± 0.12	F, $\mathscr{A}_5$	0.877 ± 0.037
0.1	C, k = 15	0.662 ± 0.047	C, k = 5	0.58 ± 0.057	C, k = 30	0.869 ± 0.061
0.1	C, k = 20	0.661 ± 0.041	F, $\mathscr{A}_6$	0.563 ± 0.099	C, k = 20	0.865 ± 0.063
0.1	F, $\mathscr{A}_2$	0.658 ± 0.04	C, k = 10	0.548 ± 0.05	C, k = 10	0.86 ± 0.06
0.1	C, k = 10	0.644 ± 0.042	C, k = 30	0.539 ± 0.048	M	0.859 ± 0.048
0.1	C, k = 5	0.631 ± 0.043	M	0.519 ± 0.038	C, k = 5	0.841 ± 0.055
0.1	C, k = 3	0.587 ± 0.034	C, k = 3	0.518 ± 0.012	C, k = 3	0.782 ± 0.088
0.1	C, k = 1	0.552 ± 0.04	C, k = 1	0.505 ± 0.01	C, k = 1	0.707 ± 0.134
0.3	C, k = 20	0.61 ± 0.061	F, $\mathscr{A}_6$	0.653 ± 0.114	F, $\mathscr{A}_1$	0.783 ± 0.113
0.3	C, k = 30	0.61 ± 0.057	F, $\mathscr{A}_5$	0.613 ± 0.095	F, $\mathscr{A}_6$	0.767 ± 0.129
0.3	F, $\mathscr{A}_1$	0.604 ± 0.067	F, $\mathscr{A}_3$	0.606 ± 0.119	F, $\mathscr{A}_3$	0.764 ± 0.111
0.3	F, $\mathscr{A}_2$	0.603 ± 0.058	F, $\mathscr{A}_2$	0.579 ± 0.1	F, $\mathscr{A}_4$	0.755 ± 0.133
0.3	F, $\mathscr{A}_4$	0.602 ± 0.066	F, $\mathscr{A}_4$	0.57 ± 0.068	F, $\mathscr{A}_2$	0.74 ± 0.16
0.3	F, $\mathscr{A}_3$	0.599 ± 0.062	F, $\mathscr{A}_1$	0.554 ± 0.067	C, k = 30	0.72 ± 0.094
0.3	C, k = 15	0.596 ± 0.051	C, k = 20	0.518 ± 0.024	F, $\mathscr{A}_5$	0.704 ± 0.075
0.3	M	0.594 ± 0.065	C, k = 3	0.513 ± 0.043	C, k = 15	0.697 ± 0.104
0.3	F, $\mathscr{A}_6$	0.594 ± 0.068	C, k = 30	0.513 ± 0.025	C, k = 10	0.686 ± 0.106
0.3	F, $\mathscr{A}_5$	0.587 ± 0.064	C, k = 10	0.507 ± 0.02	C, k = 20	0.68 ± 0.113
0.3	C, k = 10	0.579 ± 0.056	C, k = 15	0.507 ± 0.016	M	0.676 ± 0.12

(continued)

Table 8.2 (continued)

	German		Ozone		Parkinson	
Missing	Method	AUC	Method	AUC	Method	AUC
0.3	C, k = 3	0.564 ± 0.045	M	0.507 ± 0.015	C, k = 3	0.657 ± 0.097
0.3	C, k = 5	0.551 ± 0.068	C, k = 1	0.5 ± 0.0	C, k = 5	0.655 ± 0.11
0.3	C, k = 1	0.529 ± 0.025	C, k = 5	0.5 ± 0.001	C, k = 1	0.592 ± 0.071
0.5	F, $\mathscr{A}_5$	0.57 ± 0.059	F, $\mathscr{A}_2$	0.532 ± 0.041	F, $\mathscr{A}_1$	0.667 ± 0.116
0.5	F, $\mathscr{A}_6$	0.564 ± 0.069	F, $\mathscr{A}_3$	0.532 ± 0.075	F, $\mathscr{A}_2$	0.665 ± 0.133
0.5	F, $\mathscr{A}_2$	0.562 ± 0.074	F, $\mathscr{A}_1$	0.523 ± 0.025	F, $\mathscr{A}_3$	0.658 ± 0.151
0.5	F, $\mathscr{A}_3$	0.561 ± 0.064	$\mathscr{A}_4$	0.52 ± 0.027	F, $\mathscr{A}_4$	0.656 ± 0.124
0.5	M	0.56 ± 0.07	F, $\mathscr{A}_6$	0.515 0.029	F, $\mathscr{A}_6$	0.636 ± 0.136
0.5	F, $\mathscr{A}_1$	0.558 ± 0.067	F, $\mathscr{A}_5$	0.508 ± 0.036	M	0.621 ± 0.117
0.5	C, k = 30	0.557 ± 0.06	C, k = 20	0.502 ± 0.007	C, k = 5	0.617 ± 0.088
0.5	F, $\mathscr{A}_4$	0.551 ± 0.057	C, k = 3	0.501 ± 0.004	F, $\mathscr{A}_5$	0.594 ± 0.075
0.5	C, k = 10	0.546 ± 0.053	C, k = 30	0.501 ± 0.003	C, k = 15	0.588 ± 0.092
0.5	C, k = 20	0.541 ± 0.064	M	0.501 ± 0.004	C, k = 10	0.587 ± 0.104
0.5	C, k = 15	0.537 ± 0.05	C, k = 1	0.5 ± 0.0	C, k = 20	0.587 ± 0.102
0.5	C, k = 3	0.534 ± 0.037	C, k = 5	0.5 ± 0.001	C, k = 30	0.581 ± 0.151
0.5	C, k = 5	0.53 ± 0.045	C, k = 10	0.5 ± 0.001	C, k = 1	0.526 ± 0.035
0.5	C, k = 1	0.518 ± 0.021	C, k = 15	0.5 ± 0.0	C, k = 3	0.518 ± 0.117

Table 8.3 The average AUC for classification algorithms, part III

	Rethinopathy		Red wine		Spam	
Missing	Method	AUC	Method	AUC	Method	AUC
0.0	F, $\mathscr{A}_1$	0.705 ± 0.012	F, $\mathscr{A}_3$	0.806 ± 0.016	F, $\mathscr{A}_3$	0.951 ± 0.005
0.0	F, $\mathscr{A}_4$	0.703 ± 0.013	M	0.805 ± 0.009	F, $\mathscr{A}_4$	0.951 ± 0.004
0.0	F, $\mathscr{A}_3$	0.699 ± 0.017	F, $\mathscr{A}_4$	0.804 ± 0.012	M	0.95 ± 0.003
0.0	F, $\mathscr{A}_5$	0.698 ± 0.018	C, k = 30	0.802 ± 0.01	F, $\mathscr{A}_2$	0.95 ± 0.004
0.0	F, $\mathscr{A}_6$	0.696 ± 0.014	F, $\mathscr{A}_1$	0.802 ± 0.012	F, $\mathscr{A}_1$	0.949 ± 0.003
0.0	C, k = 15	0.695 ± 0.019	F, $\mathscr{A}_2$	0.801 ± 0.015	F, $\mathscr{A}_5$	0.948 ± 0.004
0.0	F, $\mathscr{A}_2$	0.693 ± 0.02	F, $\mathscr{A}_6$	0.798 ± 0.016	F, $\mathscr{A}_6$	0.948 ± 0.005
0.0	M	0.692 ± 0.014	F, $\mathscr{A}_5$	0.796 ± 0.01	C, k = 10	0.945 ± 0.003
0.0	C, k = 30	0.686 ± 0.013	C, k = 20	0.795 ± 0.01	C, k = 15	0.942 ± 0.004
0.0	C, k = 10	0.679 ± 0.02	C, k = 15	0.793 ± 0.008	C, k = 5	0.941 ± 0.005
0.0	C, k = 20	0.676 ± 0.025	C, k = 10	0.789 ± 0.013	C, k = 30	0.94 ± 0.004
0.0	C, k = 5	0.67 ± 0.019	C, k = 5	0.766 ± 0.016	C, k = 20	0.939 ± 0.005
0.0	C, k = 3	0.656 ± 0.016	C, k = 3	0.754 ± 0.008	C, k = 3	0.932 ± 0.006
0.0	C, k = 1	0.608 ± 0.013	C, k = 1	0.721 ± 0.014	C, k = 1	0.883 ± 0.008
0.05	F, $\mathscr{A}_6$	0.661 ± 0.035	F, $\mathscr{A}_2$	0.757 ± 0.029	F, $\mathscr{A}_3$	0.72 ± 0.045
0.05	F, $\mathscr{A}_2$	0.644 ± 0.063	F, $\mathscr{A}_6$	0.756 ± 0.017	F, $\mathscr{A}_4$	0.712 ± 0.05
0.05	F, $\mathscr{A}_3$	0.642 ± 0.03	F, $\mathscr{A}_1$	0.753 ± 0.024	F, $\mathscr{A}_6$	0.711 ± 0.067
0.05	F, $\mathscr{A}_1$	0.633 ± 0.036	F, $\mathscr{A}_3$	0.75 ± 0.023	F, $\mathscr{A}_2$	0.706 ± 0.058
0.05	F, $\mathscr{A}_4$	0.62 ± 0.052	F, $\mathscr{A}_4$	0.745 ± 0.022	F, $\mathscr{A}_1$	0.702 ± 0.054
0.05	M	0.609 ± 0.026	F, $\mathscr{A}_5$	0.725 ± 0.033	F, $\mathscr{A}_5$	0.702 ± 0.052
0.05	C, k = 10	0.599 ± 0.044	C, k = 30	0.687 ± 0.042	M	0.619 ± 0.015

(continued)

Table 8.3 (continued)

	Rethinopathy		Red wine		Spam	
Missing	Method	AUC	Method	AUC	Method	AUC
0.05	C, k = 5	0.591 ± 0.031	C, k = 20	0.678 ± 0.042	C, k = 15	0.614 ± 0.026
0.05	F, $\mathscr{A}_5$	0.585 ± 0.052	C, k = 10	0.67 ± 0.029	C, k = 10	0.609 ± 0.023
0.05	C, k = 15	0.582 ± 0.043	C, k = 15	0.657 ± 0.041	C, k = 20	0.608 ± 0.014
0.05	C, k = 20	0.581 ± 0.052	M	0.656 ± 0.051	C, k = 30	0.602 ± 0.033
0.05	C, k = 30	0.575 ± 0.056	C, k = 5	0.647 ± 0.052	C, k = 5	0.577 ± 0.026
0.05	C, k = 3	0.568 ± 0.03	C, k = 3	0.631 ± 0.049	C, k = 3	0.575 ± 0.025
0.05	C, k = 1	0.546 ± 0.032	C, k = 1	0.601 ± 0.041	C, k = 1	0.567 ± 0.013
0.1	F, $\mathscr{A}_5$	0.615 ± 0.039	F, $\mathscr{A}_6$	0.72 ± 0.026	F, $\mathscr{A}_2$	0.664 ± 0.051
0.1	F, $\mathscr{A}_2$	0.611 ± 0.042	F, $\mathscr{A}_3$	0.714 ± 0.051	F, $\mathscr{A}_1$	0.661 ± 0.055
0.1	F, $\mathscr{A}_1$	0.609 ± 0.049	F, $\mathscr{A}_2$	0.71 ± 0.042	F, $\mathscr{A}_5$	0.652 ± 0.045
0.1	F, $\mathscr{A}_3$	0.597 ± 0.066	F, $\mathscr{A}_4$	0.702 ± 0.05	F, $\mathscr{A}_6$	0.639 ± 0.054
0.1	F, $\mathscr{A}_4$	0.583 ± 0.052	F, $\mathscr{A}_1$	0.697 ± 0.066	F, $\mathscr{A}_4$	0.629 ± 0.048
0.1	C, k = 20	0.576 ± 0.038	F, $\mathscr{A}_5$	0.68 ± 0.047	F, $\mathscr{A}_3$	0.628 ± 0.091
0.1	F, $\mathscr{A}_6$	0.57 ± 0.065	C, k = 20	0.648 ± 0.052	M	0.584 ± 0.024
0.1	C, k = 15	0.557 ± 0.048	M	0.634 ± 0.027	C, k = 15	0.581 ± 0.031
0.1	C, k = 10	0.556 ± 0.03	C, k = 30	0.631 ± 0.04	C, k = 5	0.576 ± 0.019
0.1	C, k = 30	0.554 ± 0.038	C, k = 15	0.623 ± 0.045	C, k = 30	0.57 ± 0.031
0.1	C, k = 3	0.547 ± 0.028	C, k = 10	0.62 ± 0.053	C, k = 10	0.568 ± 0.039
0.1	M	0.543 ± 0.048	C, k = 3	0.613 ± 0.03	C, k = 20	0.565 ± 0.022
0.1	C, k = 5	0.535 ± 0.03	C, k = 5	0.606 ± 0.046	C, k = 3	0.554 ± 0.017
0.1	C, k = 1	0.526 ± 0.018	C, k = 1	0.568 ± 0.055	C, k = 1	0.538 ± 0.02

(continued)

Table 8.3 (continued)

	Rethinopathy		Red wine		Spam	
Missing	Method	AUC	Method	AUC	Method	AUC
0.3	F, $\mathscr{A}_1$	0.561 ± 0.049	F, $\mathscr{A}_3$	0.613 ± 0.061	F, $\mathscr{A}_5$	0.627 ± 0.049
0.3	F, $\mathscr{A}_6$	0.561 ± 0.054	F, $\mathscr{A}_1$	0.608 ± 0.075	F, $\mathscr{A}_1$	0.6 ± 0.038
0.3	F, $\mathscr{A}_3$	0.558 ± 0.034	F, $\mathscr{A}_4$	0.607 ± 0.06	F, $\mathscr{A}_3$	0.599 ± 0.048
0.3	F, $\mathscr{A}_4$	0.55 ± 0.036	F, $\mathscr{A}_2$	0.604 ± 0.045	F, $\mathscr{A}_6$	0.591 ± 0.061
0.3	F, $\mathscr{A}_2$	0.546 ± 0.035	F, $\mathscr{A}_6$	0.601 ± 0.052	F, $\mathscr{A}_4$	0.589 ± 0.045
0.3	F, $\mathscr{A}_5$	0.544 ± 0.039	F, $\mathscr{A}_5$	0.584 ± 0.074	F, $\mathscr{A}_2$	0.58 ± 0.046
0.3	C, k = 30	0.524 ± 0.022	M	0.558 ± 0.03	M	0.575 ± 0.05
0.3	M	0.524 ± 0.031	C, k = 3	0.552 ± 0.05	C, k = 5	0.565 ± 0.043
0.3	C, k = 10	0.521 ± 0.019	C, k = 5	0.549 ± 0.033	C, k = 10	0.546 ± 0.031
0.3	C, k = 15	0.521 ± 0.024	C, k = 30	0.545 ± 0.036	C, k = 20	0.543 ± 0.042
0.3	C, k = 20	0.52 ± 0.02	C, k = 15	0.543 ± 0.05	C, k = 15	0.542 ± 0.033
0.3	C, k = 3	0.513 ± 0.011	C, k = 10	0.54 ± 0.035	C, k = 30	0.537 ± 0.019
0.3	C, k = 5	0.5 ± 0.024	C, k = 1	0.535 ± 0.029	C, k = 3	0.531 ± 0.029
0.3	C, k = 1	0.496 ± 0.01	C, k = 20	0.518 ± 0.044	C, k = 1	0.528 ± 0.024
0.5	F, $\mathscr{A}_2$	0.548 ± 0.042	F, $\mathscr{A}_2$	0.582 ± 0.063	F, $\mathscr{A}_6$	0.589 ± 0.108
0.5	F, $\mathscr{A}_1$	0.524 ± 0.029	F, $\mathscr{A}_1$	0.571 ± 0.061	F, $\mathscr{A}_2$	0.582 ± 0.036
0.5	C, k = 10	0.522 ± 0.017	F, $\mathscr{A}_3$	0.57 ± 0.045	F, $\mathscr{A}_4$	0.566 ± 0.06
0.5	F, $\mathscr{A}_3$	0.521 ± 0.023	F, $\mathscr{A}_5$	0.554 ± 0.057	F, $\mathscr{A}_5$	0.566 ± 0.075
0.5	C, k = 5	0.516 ± 0.013	F, $\mathscr{A}_6$	0.545 ± 0.053	M	0.558 ± 0.062
0.5	C, k = 30	0.514 ± 0.023	F, $\mathscr{A}_4$	0.544 ± 0.061	C, k = 20	0.547 ± 0.029
0.5	C, k = 3	0.511 ± 0.013	C, k = 5	0.529 ± 0.022	C, k = 5	0.542 ± 0.044

(continued)

Table 8.3 (continued)

	Rethinopathy		Red wine		Spam	
Missing	Method	AUC	Method	AUC	Method	AUC
0.5	C, k = 15	0.511 ± 0.021	M	0.528 ± 0.036	F, $\mathscr{A}_3$	0.542 ± 0.075
0.5	F, $\mathscr{A}_4$	0.51 ± 0.033	C, k = 15	0.527 ± 0.038	C, k = 15	0.536 ± 0.05
0.5	F, $\mathscr{A}_5$	0.507 ± 0.028	C, k = 20	0.527 ± 0.039	F, $\mathscr{A}_1$	0.533 ± 0.075
0.5	M	0.504 ± 0.027	C, k = 30	0.525 ± 0.032	C, k = 10	0.521 ± 0.021
0.5	C, k = 1	0.501 ± 0.01	C, k = 3	0.515 ± 0.021	C, k = 30	0.501 ± 0.035
0.5	C, k = 20	0.499 ± 0.024	C, k = 1	0.512 ± 0.016	C, k = 1	0.498 ± 0.028
0.5	F, $\mathscr{A}_6$	0.496 ± 0.032	C, k = 10	0.509 ± 0.025	C, k = 3	0.474 ± 0.068

Table 8.4 The average ACC for DNA microarrays method, Colon data set

NoClassHoriz	NoClassVert	Method	k-NN	ACC	SN	SP
10	50	F, $\mathscr{A}_4$	9	0.719 ± 0.03	0.72 ± 0.031	0.718 ± 0.032
10	50	F, $\mathscr{A}_2$	9	0.714 ± 0.044	0.714 ± 0.044	0.714 ± 0.047
10	50	F, $\mathscr{A}_3$	9	0.713 ± 0.036	0.713 ± 0.039	0.714 ± 0.033
10	50	F, $\mathscr{A}_1$	9	0.712 ± 0.043	0.714 ± 0.045	0.709 ± 0.043
10	50	F, $\mathscr{A}_6$	9	0.681 ± 0.04	0.68 ± 0.039	0.684 ± 0.045
10	50	F, $\mathscr{A}_6$	7	0.678 ± 0.048	0.68 ± 0.046	0.675 ± 0.058
10	50	F, $\mathscr{A}_3$	7	0.668 ± 0.036	0.669 ± 0.037	0.666 ± 0.037
10	50	F, $\mathscr{A}_4$	7	0.663 ± 0.039	0.659 ± 0.042	0.67 ± 0.039
10	50	F, $\mathscr{A}_1$	7	0.661 ± 0.042	0.663 ± 0.041	0.659 ± 0.048
10	50	F, $\mathscr{A}_2$	7	0.642 ± 0.048	0.644 ± 0.045	0.639 ± 0.06
10	50	F, $\mathscr{A}_4$	5	0.64 ± 0.049	0.644 ± 0.075	0.632 ± 0.081
10	50	F, $\mathscr{A}_5$	5	0.64 ± 0.025	0.963 ± 0.031	0.055 ± 0.05

(continued)

Table 8.4 (continued)

NoClassHoriz	NoClassVert	Method	k-NN	ACC	SN	SP
10	50	M	7	0.632 ± 0.055	0.634 ± 0.07	0.63 ± 0.061
10	50	F, $\mathscr{A}_2$	5	0.626 ± 0.046	0.633 ± 0.043	0.614 ± 0.077
10	50	F, $\mathscr{A}_5$	9	0.625 ± 0.046	0.644 ± 0.078	0.591 ± 0.055
10	50	M	9	0.62 ± 0.038	0.635 ± 0.055	0.593 ± 0.062
10	50	M	3	0.613 ± 0.039	0.625 ± 0.059	0.591 ± 0.063
10	50	M	5	0.612 ± 0.041	0.606 ± 0.052	0.623 ± 0.072
10	50	F, $\mathscr{A}_5$	7	0.598 ± 0.07	0.594 ± 0.146	0.607 ± 0.112
10	50	F, $\mathscr{A}_3$	5	0.594 ± 0.059	0.624 ± 0.078	0.541 ± 0.076
10	50	F, $\mathscr{A}_1$	5	0.58 ± 0.058	0.591 ± 0.074	0.559 ± 0.102
10	50	F, $\mathscr{A}_6$	5	0.569 ± 0.065	0.578 ± 0.066	0.552 ± 0.082
10	50	F, $\mathscr{A}_3$	3	0.568 ± 0.06	0.631 ± 0.075	0.452 ± 0.124
10	50	F, $\mathscr{A}_4$	3	0.558 ± 0.048	0.638 ± 0.05	0.414 ± 0.094
10	50	F, $\mathscr{A}_1$	3	0.557 ± 0.06	0.629 ± 0.109	0.427 ± 0.103
10	50	F, $\mathscr{A}_2$	3	0.551 ± 0.054	0.585 ± 0.09	0.489 ± 0.093
10	50	F, $\mathscr{A}_6$	3	0.54 ± 0.057	0.601 ± 0.089	0.43 ± 0.106
10	50	F, $\mathscr{A}_5$	3	0.463 ± 0.123	0.333 ± 0.449	0.7 ± 0.47
10	10	F, $\mathscr{A}_3$	9	0.668 ± 0.046	0.665 ± 0.044	0.673 ± 0.052
10	10	F, $\mathscr{A}_6$	7	0.651 ± 0.062	0.654 ± 0.06	0.645 ± 0.07
10	10	F, $\mathscr{A}_6$	9	0.648 ± 0.048	0.649 ± 0.049	0.648 ± 0.049
10	10	F, $\mathscr{A}_2$	7	0.647 ± 0.048	0.649 ± 0.046	0.643 ± 0.054
10	10	F, $\mathscr{A}_4$	7	0.644 ± 0.045	0.648 ± 0.044	0.639 ± 0.05
10	10	F, $\mathscr{A}_4$	9	0.644 ± 0.045	0.645 ± 0.043	0.643 ± 0.052

(continued)

Table 8.4 (continued)

NoClassHoriz	NoClassVert	Method	k-NN	ACC	SN	SP
10	10	F, $\mathscr{A}_1$	7	0.639 ± 0.062	0.64 ± 0.064	0.636 ± 0.064
10	10	F, $\mathscr{A}_2$	9	0.639 ± 0.038	0.642 ± 0.036	0.632 ± 0.044
10	10	F, $\mathscr{A}_1$	9	0.638 ± 0.044	0.641 ± 0.042	0.632 ± 0.053
10	10	F, $\mathscr{A}_3$	7	0.636 ± 0.043	0.643 ± 0.039	0.625 ± 0.055
10	10	F, $\mathscr{A}_5$	9	0.631 ± 0.047	0.633 ± 0.045	0.63 ± 0.056
10	10	F, $\mathscr{A}_4$	5	0.617 ± 0.05	0.619 ± 0.048	0.614 ± 0.06
10	10	F, $\mathscr{A}_1$	5	0.61 ± 0.044	0.615 ± 0.042	0.602 ± 0.051
10	10	F, $\mathscr{A}_3$	5	0.604 ± 0.049	0.606 ± 0.048	0.6 ± 0.058
10	10	M	7	0.602 ± 0.045	0.607 ± 0.069	0.591 ± 0.094
10	10	F, $\mathscr{A}_5$	2	0.602 ± 0.059	0.614 ± 0.058	0.582 ± 0.065
10	10	F, $\mathscr{A}_5$	7	0.594 ± 0.072	0.608 ± 0.073	0.568 ± 0.092
10	10	F, $\mathscr{A}_6$	5	0.592 ± 0.035	0.595 ± 0.039	0.586 ± 0.033
10	10	F, $\mathscr{A}_4$	3	0.589 ± 0.066	0.595 ± 0.067	0.577 ± 0.08
10	10	F, $\mathscr{A}_6$	3	0.587 ± 0.063	0.601 ± 0.067	0.561 ± 0.065
10	10	F, $\mathscr{A}_1$	3	0.586 ± 0.054	0.587 ± 0.056	0.584 ± 0.059
10	10	F, $\mathscr{A}_3$	3	0.586 ± 0.052	0.594 ± 0.055	0.573 ± 0.067
10	10	M	9	0.577 ± 0.053	0.595 ± 0.072	0.543 ± 0.107
10	10	F, $\mathscr{A}_5$	3	0.565 ± 0.056	0.638 ± 0.069	0.434 ± 0.084
10	10	M	5	0.556 ± 0.051	0.555 ± 0.078	0.559 ± 0.142
10	10	F, $\mathscr{A}_5$	5	0.555 ± 0.059	0.571 ± 0.087	0.525 ± 0.119
10	10	F, $\mathscr{A}_2$	3	0.552 ± 0.062	0.554 ± 0.062	0.548 ± 0.077
10	10	M	3	0.55 ± 0.068	0.564 ± 0.075	0.525 ± 0.165

(continued)

Table 8.4 (continued)

NoClassHoriz	NoClassVert	Method	k-NN	ACC	SN	SP
10	5	F, $\mathscr{A}_3$	9	0.624 ± 0.04	0.628 ± 0.038	0.618 ± 0.048
10	5	F, $\mathscr{A}_4$	9	0.624 ± 0.047	0.625 ± 0.047	0.623 ± 0.051
10	5	F, $\mathscr{A}_6$	9	0.623 ± 0.049	0.622 ± 0.047	0.625 ± 0.055
10	5	F, $\mathscr{A}_1$	9	0.622 ± 0.043	0.624 ± 0.042	0.618 ± 0.05
10	5	F, $\mathscr{A}_6$	7	0.62 ± 0.046	0.624 ± 0.042	0.614 ± 0.058
10	5	F, $\mathscr{A}_4$	7	0.619 ± 0.047	0.62 ± 0.048	0.618 ± 0.05
10	5	F, $\mathscr{A}_1$	5	0.614 ± 0.054	0.617 ± 0.056	0.607 ± 0.058
10	5	F, $\mathscr{A}_3$	7	0.612 ± 0.062	0.613 ± 0.063	0.611 ± 0.062
10	5	F, $\mathscr{A}_2$	7	0.609 ± 0.053	0.614 ± 0.05	0.6 ± 0.064
10	5	F, $\mathscr{A}_1$	7	0.608 ± 0.042	0.615 ± 0.043	0.595 ± 0.044
10	5	F, $\mathscr{A}_6$	3	0.607 ± 0.059	0.615 ± 0.058	0.593 ± 0.07
10	5	F, $\mathscr{A}_2$	5	0.604 ± 0.07	0.604 ± 0.069	0.605 ± 0.074
10	5	F, $\mathscr{A}_2$	9	0.604 ± 0.051	0.609 ± 0.051	0.595 ± 0.055
10	5	F, $\mathscr{A}_4$	5	0.595 ± 0.05	0.604 ± 0.049	0.58 ± 0.057
10	5	F, $\mathscr{A}_1$	3	0.594 ± 0.052	0.6 ± 0.053	0.582 ± 0.06
10	5	F, $\mathscr{A}_3$	5	0.589 ± 0.053	0.595 ± 0.05	0.577 ± 0.061
10	5	F, $\mathscr{A}_4$	3	0.584 ± 0.067	0.593 ± 0.069	0.568 ± 0.074
10	5	F, $\mathscr{A}_3$	3	0.57 ± 0.038	0.581 ± 0.043	0.55 ± 0.041
10	5	F, $\mathscr{A}_6$	5	0.569 ± 0.047	0.576 ± 0.049	0.557 ± 0.051
10	5	M	3	0.565 ± 0.06	0.594 ± 0.099	0.511 ± 0.101
10	5	F, $\mathscr{A}_5$	7	0.565 ± 0.058	0.569 ± 0.065	0.559 ± 0.053
10	5	M	9	0.564 ± 0.052	0.568 ± 0.09	0.557 ± 0.082

(continued)

Table 8.4 (continued)

NoClassHoriz	NoClassVert	Method	k-NN	ACC	SN	SP
10	5	F, $\mathscr{A}_5$	9	0.563 ± 0.068	0.568 ± 0.067	0.555 ± 0.08
10	5	F, $\mathscr{A}_5$	5	0.561 ± 0.069	0.57 ± 0.081	0.545 ± 0.075
10	5	F, $\mathscr{A}_2$	3	0.559 ± 0.072	0.561 ± 0.072	0.555 ± 0.077
10	5	M	5	0.547 ± 0.048	0.541 ± 0.08	0.557 ± 0.135
10	5	F, $\mathscr{A}_5$	3	0.544 ± 0.065	0.58 ± 0.115	0.477 ± 0.106
10	5	M	7	0.542 ± 0.046	0.539 ± 0.083	0.548 ± 0.114

Table 8.5 The average ACC for DNA microarrays method, Leukemia data set

NoClassHoriz	NoClassVert	Method	k-NN	ACC	SN	SP
10	50	F, $\mathscr{A}_2$	9	0.878 ± 0.034	0.88 ± 0.034	0.876 ± 0.036
10	50	F, $\mathscr{A}_1$	9	0.876 ± 0.032	0.877 ± 0.032	0.876 ± 0.034
10	50	F, $\mathscr{A}_3$	9	0.874 ± 0.022	0.876 ± 0.021	0.87 ± 0.026
10	50	F, $\mathscr{A}_4$	9	0.86 ± 0.03	0.862 ± 0.03	0.856 ± 0.03
10	50	F, $\mathscr{A}_6$	9	0.86 ± 0.028	0.862 ± 0.03	0.858 ± 0.024
10	50	F, $\mathscr{A}_1$	7	0.856 ± 0.035	0.856 ± 0.035	0.854 ± 0.04
10	50	F, $\mathscr{A}_4$	7	0.849 ± 0.052	0.851 ± 0.053	0.846 ± 0.054
10	50	F, $\mathscr{A}_3$	7	0.844 ± 0.039	0.846 ± 0.039	0.842 ± 0.04
10	50	F, $\mathscr{A}_6$	7	0.835 ± 0.033	0.837 ± 0.033	0.832 ± 0.033
10	50	F, $\mathscr{A}_2$	7	0.825 ± 0.028	0.828 ± 0.029	0.82 ± 0.03
10	50	F, $\mathscr{A}_3$	5	0.824 ± 0.031	0.826 ± 0.036	0.82 ± 0.04
10	50	F, $\mathscr{A}_1$	5	0.819 ± 0.035	0.824 ± 0.038	0.81 ± 0.034
10	50	F, $\mathscr{A}_5$	9	0.811 ± 0.045	0.834 ± 0.065	0.768 ± 0.068

(continued)

Table 8.5 (continued)

NoClassHoriz	NoClassVert	Method	k-NN	ACC	SN	SP
10	50	F, $\mathscr{A}_4$	5	0.806 ± 0.044	0.807 ± 0.043	0.804 ± 0.048
10	50	M	3	0.799 ± 0.024	0.798 ± 0.043	0.802 ± 0.033
10	50	F, $\mathscr{A}_2$	5	0.797 ± 0.032	0.797 ± 0.036	0.798 ± 0.027
10	50	M	7	0.795 ± 0.022	0.798 ± 0.028	0.79 ± 0.039
10	50	M	9	0.793 ± 0.022	0.797 ± 0.03	0.786 ± 0.033
10	50	M	5	0.783 ± 0.02	0.781 ± 0.029	0.788 ± 0.035
10	50	F, $\mathscr{A}_6$	5	0.783 ± 0.041	0.784 ± 0.044	0.78 ± 0.048
10	50	F, $\mathscr{A}_5$	7	0.768 ± 0.034	0.902 ± 0.049	0.516 ± 0.065
10	50	F, $\mathscr{A}_1$	3	0.714 ± 0.056	0.719 ± 0.057	0.704 ± 0.061
10	50	F, $\mathscr{A}_6$	3	0.708 ± 0.047	0.703 ± 0.047	0.716 ± 0.053
10	50	F, $\mathscr{A}_2$	3	0.703 ± 0.046	0.698 ± 0.05	0.714 ± 0.055
10	50	F, $\mathscr{A}_3$	3	0.7 ± 0.039	0.695 ± 0.045	0.71 ± 0.052
10	50	F, $\mathscr{A}_4$	3	0.689 ± 0.042	0.682 ± 0.042	0.702 ± 0.051
10	50	F, $\mathscr{A}_5$	5	0.666 ± 0.016	0.999 ± 0.005	0.04 ± 0.043
10	50	F, $\mathscr{A}_3$	5	0.653 ± 0.0	1.0 ± 0.0	0.0 ± 0.0
10	10	F, $\mathscr{A}_4$	9	0.822 ± 0.029	0.823 ± 0.031	0.82 ± 0.028
10	10	F, $\mathscr{A}_2$	9	0.817 ± 0.04	0.819 ± 0.04	0.814 ± 0.04
10	10	F, $\mathscr{A}_2$	7	0.814 ± 0.038	0.816 ± 0.039	0.81 ± 0.039
10	10	F, $\mathscr{A}_1$	7	0.813 ± 0.037	0.815 ± 0.037	0.81 ± 0.036
10	10	F, $\mathscr{A}_3$	7	0.808 ± 0.048	0.81 ± 0.048	0.806 ± 0.051
10	10	F, $\mathscr{A}_4$	7	0.808 ± 0.051	0.811 ± 0.052	0.804 ± 0.05
10	10	F, $\mathscr{A}_6$	7	0.808 ± 0.042	0.809 ± 0.044	0.806 ± 0.04

(continued)

Table 8.5 (continued)

NoClassHoriz	NoClassVert	Method	k-NN	ACC	SN	SP
10	10	F, $\mathscr{A}_3$	9	0.808 ± 0.039	0.811 ± 0.038	0.802 ± 0.04
10	10	F, $\mathscr{A}_6$	9	0.804 ± 0.038	0.805 ± 0.038	0.802 ± 0.038
10	10	F, $\mathscr{A}_1$	9	0.801 ± 0.038	0.803 ± 0.038	0.798 ± 0.038
10	10	F, $\mathscr{A}_3$	5	0.788 ± 0.027	0.789 ± 0.028	0.786 ± 0.027
10	10	F, $\mathscr{A}_4$	5	0.785 ± 0.045	0.785 ± 0.044	0.786 ± 0.047
10	10	M	9	0.782 ± 0.032	0.788 ± 0.058	0.77 ± 0.052
10	10	M	7	0.778 ± 0.041	0.777 ± 0.065	0.78 ± 0.051
10	10	F, $\mathscr{A}_1$	5	0.776 ± 0.038	0.778 ± 0.039	0.774 ± 0.037
10	10	F, $\mathscr{A}_2$	5	0.776 ± 0.035	0.778 ± 0.036	0.774 ± 0.033
10	10	F, $\mathscr{A}_6$	5	0.776 ± 0.046	0.778 ± 0.046	0.772 ± 0.045
10	10	F, $\mathscr{A}_5$	9	0.762 ± 0.045	0.765 ± 0.048	0.756 ± 0.045
10	10	M	3	0.753 ± 0.047	0.743 ± 0.072	0.772 ± 0.069
10	10	M	5	0.749 ± 0.028	0.735 ± 0.055	0.774 ± 0.057
10	10	F, $\mathscr{A}_3$	3	0.744 ± 0.058	0.744 ± 0.06	0.744 ± 0.057
10	10	F, $\mathscr{A}_4$	3	0.742 ± 0.051	0.74 ± 0.048	0.746 ± 0.058
10	10	F, $\mathscr{A}_5$	5	0.74 ± 0.057	0.76 ± 0.094	0.702 ± 0.068
10	10	F, $\mathscr{A}_1$	3	0.738 ± 0.04	0.738 ± 0.04	0.738 ± 0.042
10	10	F, $\mathscr{A}_2$	3	0.731 ± 0.038	0.732 ± 0.037	0.728 ± 0.048
10	10	F, $\mathscr{A}_6$	3	0.726 ± 0.047	0.726 ± 0.046	0.726 ± 0.052
10	10	F, $\mathscr{A}_5$	7	0.722 ± 0.051	0.717 ± 0.057	0.732 ± 0.061
10	10	F, $\mathscr{A}_5$	3	0.679 ± 0.046	0.81 ± 0.053	0.434 ± 0.106
10	5	F, $\mathscr{A}_4$	9	0.796 ± 0.045	0.795 ± 0.043	0.798 ± 0.048

(continued)

Table 8.5 (continued)

NoClassHoriz	NoClassVert	Method	k-NN	ACC	SN	SP
10	5	F, $\mathscr{A}_3$	9	0.781 ± 0.036	0.782 ± 0.037	0.78 ± 0.036
10	5	F, $\mathscr{A}_2$	5	0.78 ± 0.039	0.781 ± 0.04	0.778 ± 0.038
10	5	F, $\mathscr{A}_3$	7	0.78 ± 0.047	0.781 ± 0.047	0.778 ± 0.046
10	5	F, $\mathscr{A}_4$	7	0.779 ± 0.036	0.78 ± 0.037	0.778 ± 0.035
10	5	F, $\mathscr{A}_3$	5	0.778 ± 0.04	0.779 ± 0.039	0.778 ± 0.042
10	5	F, $\mathscr{A}_6$	7	0.778 ± 0.036	0.779 ± 0.036	0.778 ± 0.038
10	5	F, $\mathscr{A}_2$	9	0.778 ± 0.046	0.779 ± 0.045	0.776 ± 0.048
10	5	F, $\mathscr{A}_2$	7	0.774 ± 0.049	0.773 ± 0.049	0.774 ± 0.049
10	5	F, $\mathscr{A}_6$	9	0.77 ± 0.042	0.77 ± 0.043	0.77 ± 0.041
10	5	F, $\mathscr{A}_4$	5	0.769 ± 0.046	0.769 ± 0.048	0.768 ± 0.044
10	5	F, $\mathscr{A}_1$	7	0.769 ± 0.048	0.768 ± 0.05	0.772 ± 0.045
10	5	F, $\mathscr{A}_1$	9	0.762 ± 0.033	0.763 ± 0.033	0.76 ± 0.032
10	5	M	7	0.752 ± 0.036	0.748 ± 0.057	0.76 ± 0.055
10	5	M	9	0.751 ± 0.034	0.743 ± 0.044	0.766 ± 0.081
10	5	F, $\mathscr{A}_2$	3	0.751 ± 0.052	0.752 ± 0.053	0.75 ± 0.052
10	5	F, $\mathscr{A}_6$	5	0.75 ± 0.042	0.75 ± 0.042	0.75 ± 0.043
10	5	F, $\mathscr{A}_1$	5	0.749 ± 0.04	0.749 ± 0.04	0.748 ± 0.039
10	5	F, $\mathscr{A}_4$	3	0.742 ± 0.049	0.743 ± 0.051	0.742 ± 0.048
10	5	F, $\mathscr{A}_6$	3	0.735 ± 0.046	0.734 ± 0.043	0.736 ± 0.053
10	5	M	5	0.732 ± 0.042	0.736 ± 0.058	0.724 ± 0.069
10	5	M	3	0.731 ± 0.05	0.729 ± 0.069	0.734 ± 0.085
10	5	F, $\mathscr{A}_3$	3	0.731 ± 0.034	0.73 ± 0.034	0.732 ± 0.035

(continued)

Table 8.5 (continued)

NoClassHoriz	NoClassVert	Method	k-NN	ACC	SN	SP
10	5	F, $\mathscr{A}_1$	3	0.73 ± 0.053	0.729 ± 0.053	0.732 ± 0.057
10	5	F, $\mathscr{A}_5$	9	0.724 ± 0.049	0.723 ± 0.048	0.724 ± 0.053
10	5	F, $\mathscr{A}_5$	7	0.717 ± 0.055	0.717 ± 0.057	0.716 ± 0.055
10	5	F, $\mathscr{A}_5$	5	0.679 ± 0.056	0.68 ± 0.055	0.678 ± 0.072
10	5	F, $\mathscr{A}_5$	3	0.647 ± 0.046	0.647 ± 0.07	0.648 ± 0.077

Table 8.6 The average ACC for DNA microarrays method, Lymphoma data set

NoClassHoriz	NoClassVert	Method	k-NN	ACC	SN	SP
10	50	F, $\mathscr{A}_3$	9	0.851 ± 0.028	0.85 ± 0.026	0.852 ± 0.032
10	50	F, $\mathscr{A}_1$	9	0.84 ± 0.039	0.843 ± 0.036	0.837 ± 0.045
10	50	F, $\mathscr{A}_4$	9	0.836 ± 0.034	0.835 ± 0.033	0.838 ± 0.036
10	50	F, $\mathscr{A}_2$	7	0.835 ± 0.029	0.837 ± 0.024	0.833 ± 0.036
10	50	F, $\mathscr{A}_6$	9	0.831 ± 0.036	0.833 ± 0.032	0.829 ± 0.04
10	50	F, $\mathscr{A}_4$	7	0.829 ± 0.045	0.83 ± 0.04	0.827 ± 0.051
10	50	F, $\mathscr{A}_6$	7	0.828 ± 0.041	0.828 ± 0.036	0.827 ± 0.047
10	50	F, $\mathscr{A}_6$	7	0.826 ± 0.042	0.828 ± 0.039	0.823 ± 0.047
10	50	F, $\mathscr{A}_1$	7	0.824 ± 0.04	0.828 ± 0.036	0.821 ± 0.047
10	50	F, $\mathscr{A}_2$	9	0.821 ± 0.042	0.822 ± 0.04	0.821 ± 0.045
10	50	F, $\mathscr{A}_5$	9	0.801 ± 0.051	0.804 ± 0.046	0.798 ± 0.061
10	50	F, $\mathscr{A}_5$	7	0.786 ± 0.048	0.785 ± 0.067	0.787 ± 0.047
10	50	F, $\mathscr{A}_4$	5	0.785 ± 0.044	0.787 ± 0.047	0.783 ± 0.044
10	50	M	7	0.781 ± 0.043	0.776 ± 0.071	0.785 ± 0.043

(continued)

Table 8.6 (continued)

NoClassHoriz	NoClassVert	Method	k-NN	ACC	SN	SP
10	50	M	3	0.78 ± 0.032	0.798 ± 0.068	0.763 ± 0.049
10	50	F, $\mathscr{A}_1$	5	0.778 ± 0.041	0.78 ± 0.043	0.775 ± 0.041
10	50	F, $\mathscr{A}_3$	5	0.778 ± 0.048	0.783 ± 0.049	0.773 ± 0.05
10	50	F, $\mathscr{A}_2$	5	0.771 ± 0.038	0.778 ± 0.037	0.765 ± 0.041
10	50	M	5	0.766 ± 0.032	0.787 ± 0.06	0.746 ± 0.047
10	50	M	9	0.765 ± 0.037	0.77 ± 0.069	0.76 ± 0.05
10	50	F, $\mathscr{A}_6$	5	0.757 ± 0.04	0.761 ± 0.041	0.754 ± 0.043
10	50	F, $\mathscr{A}_5$	5	0.724 ± 0.055	0.722 ± 0.096	0.727 ± 0.068
10	50	F, $\mathscr{A}_2$	3	0.562 ± 0.049	0.57 ± 0.154	0.554 ± 0.169
10	50	F, $\mathscr{A}_3$	3	0.555 ± 0.064	0.576 ± 0.202	0.535 ± 0.179
10	50	F, $\mathscr{A}_5$	3	0.549 ± 0.078	0.343 ± 0.113	0.746 ± 0.089
10	50	F, $\mathscr{A}_4$	3	0.545 ± 0.074	0.546 ± 0.159	0.544 ± 0.21
10	50	F, $\mathscr{A}_6$	3	0.54 ± 0.079	0.511 ± 0.167	0.569 ± 0.226
10	50	F, $\mathscr{A}_1$	3	0.536 ± 0.06	0.522 ± 0.149	0.55 ± 0.2
10	10	F, $\mathscr{A}_4$	7	0.796 ± 0.049	0.798 ± 0.047	0.794 ± 0.053
10	10	F, $\mathscr{A}_1$	9	0.79 ± 0.058	0.796 ± 0.055	0.785 ± 0.062
10	10	F, $\mathscr{A}_6$	9	0.79 ± 0.049	0.793 ± 0.049	0.788 ± 0.052
10	10	F, $\mathscr{A}_4$	5	0.781 ± 0.05	0.785 ± 0.046	0.777 ± 0.056
10	10	F, $\mathscr{A}_4$	9	0.779 ± 0.039	0.783 ± 0.037	0.775 ± 0.044
10	10	F, $\mathscr{A}_2$	7	0.773 ± 0.045	0.776 ± 0.045	0.771 ± 0.048
10	10	F, $\mathscr{A}_1$	7	0.768 ± 0.049	0.774 ± 0.052	0.763 ± 0.049
10	10	F, $\mathscr{A}_3$	7	0.767 ± 0.051	0.77 ± 0.053	0.765 ± 0.051

(continued)

Table 8.6 (continued)

NoClassHoriz	NoClassVert	Method	k-NN	ACC	SN	SP
10	10	F, $\mathscr{A}_3$	5	0.766 ± 0.047	0.772 ± 0.047	0.76 ± 0.05
10	10	F, $\mathscr{A}_2$	9	0.766 ± 0.046	0.776 ± 0.045	0.756 ± 0.049
10	10	F, $\mathscr{A}_3$	9	0.766 ± 0.051	0.77 ± 0.051	0.763 ± 0.053
10	10	F, $\mathscr{A}_6$	7	0.759 ± 0.044	0.761 ± 0.043	0.756 ± 0.047
10	10	F, $\mathscr{A}_2$	5	0.751 ± 0.055	0.757 ± 0.062	0.746 ± 0.05
10	10	F, $\mathscr{A}_6$	5	0.748 ± 0.051	0.75 ± 0.054	0.746 ± 0.05
10	10	F, $\mathscr{A}_4$	3	0.745 ± 0.051	0.752 ± 0.055	0.738 ± 0.051
10	10	F, $\mathscr{A}_1$	5	0.745 ± 0.047	0.75 ± 0.051	0.74 ± 0.047
10	10	M	9	0.736 ± 0.042	0.743 ± 0.089	0.729 ± 0.095
10	10	M	7	0.731 ± 0.05	0.75 ± 0.11	0.712 ± 0.075
10	10	M	5	0.729 ± 0.059	0.702 ± 0.113	0.754 ± 0.086
10	10	F, $\mathscr{A}_6$	3	0.727 ± 0.058	0.724 ± 0.059	0.729 ± 0.061
10	10	F, $\mathscr{A}_1$	3	0.721 ± 0.055	0.724 ± 0.055	0.719 ± 0.059
10	10	F, $\mathscr{A}_5$	9	0.72 ± 0.047	0.724 ± 0.051	0.717 ± 0.046
10	10	F, $\mathscr{A}_2$	3	0.718 ± 0.053	0.726 ± 0.058	0.71 ± 0.051
10	10	F, $\mathscr{A}_3$	3	0.718 ± 0.071	0.72 ± 0.071	0.717 ± 0.072
10	10	F, $\mathscr{A}_5$	7	0.717 ± 0.065	0.717 ± 0.071	0.717 ± 0.061
10	10	M	3	0.715 ± 0.049	0.702 ± 0.096	0.727 ± 0.102
10	10	F, $\mathscr{A}_5$	5	0.714 ± 0.054	0.717 ± 0.062	0.71 ± 0.051
10	10	F, $\mathscr{A}_5$	3	0.636 ± 0.066	0.65 ± 0.094	0.623 ± 0.079
10	5	F, $\mathscr{A}_6$	7	0.756 ± 0.066	0.763 ± 0.07	0.75 ± 0.065
10	5	F, $\mathscr{A}_2$	7	0.755 ± 0.04	0.763 ± 0.039	0.748 ± 0.044

(continued)

Table 8.6 (continued)

NoClassHoriz	NoClassVert	Method	k-NN	ACC	SN	SP
10	5	F, $\mathscr{A}_1$	9	0.754 ± 0.039	0.759 ± 0.041	0.75 ± 0.041
10	5	F, $\mathscr{A}_4$	5	0.753 ± 0.042	0.757 ± 0.043	0.75 ± 0.043
10	5	F, $\mathscr{A}_3$	9	0.748 ± 0.058	0.748 ± 0.056	0.748 ± 0.061
10	5	F, $\mathscr{A}_1$	7	0.745 ± 0.064	0.746 ± 0.065	0.744 ± 0.064
10	5	F, $\mathscr{A}_3$	7	0.741 ± 0.046	0.741 ± 0.046	0.742 ± 0.048
10	5	F, $\mathscr{A}_6$	9	0.741 ± 0.063	0.75 ± 0.064	0.733 ± 0.063
10	5	F, $\mathscr{A}_2$	9	0.739 ± 0.065	0.743 ± 0.072	0.735 ± 0.061
10	5	F, $\mathscr{A}_3$	5	0.737 ± 0.065	0.743 ± 0.063	0.731 ± 0.068
10	5	F, $\mathscr{A}_4$	7	0.737 ± 0.05	0.743 ± 0.053	0.731 ± 0.05
10	5	F, $\mathscr{A}_4$	9	0.736 ± 0.052	0.737 ± 0.055	0.735 ± 0.051
10	5	F, $\mathscr{A}_2$	5	0.734 ± 0.056	0.735 ± 0.058	0.733 ± 0.056
10	5	F, $\mathscr{A}_1$	5	0.726 ± 0.05	0.726 ± 0.053	0.725 ± 0.049
10	5	F, $\mathscr{A}_6$	5	0.722 ± 0.048	0.728 ± 0.049	0.717 ± 0.05
10	5	F, $\mathscr{A}_6$	3	0.721 ± 0.046	0.726 ± 0.049	0.717 ± 0.046
10	5	F, $\mathscr{A}_4$	3	0.716 ± 0.07	0.724 ± 0.072	0.708 ± 0.07
10	5	F, $\mathscr{A}_3$	3	0.715 ± 0.045	0.72 ± 0.046	0.71 ± 0.048
10	5	F, $\mathscr{A}_1$	3	0.712 ± 0.053	0.713 ± 0.055	0.71 ± 0.053
10	5	F, $\mathscr{A}_2$	3	0.701 ± 0.058	0.702 ± 0.06	0.7 ± 0.057
10	5	M	9	0.695 ± 0.058	0.741 ± 0.101	0.65 ± 0.094
10	5	M	5	0.689 ± 0.065	0.689 ± 0.12	0.69 ± 0.092
10	5	M	7	0.688 ± 0.058	0.689 ± 0.133	0.688 ± 0.092
10	5	F, $\mathscr{A}_5$	7	0.688 ± 0.064	0.693 ± 0.065	0.683 ± 0.064

(continued)

Table 8.6 (continued)

NoClassHoriz	NoClassVert	Method	k-NN	ACC	SN	SP
10	5	F, $\mathscr{A}_5$	5	0.682 ± 0.057	0.685 ± 0.061	0.679 ± 0.058
10	5	F, $\mathscr{A}_5$	9	0.677 ± 0.061	0.683 ± 0.06	0.671 ± 0.065
10	5	M	3	0.666 ± 0.062	0.68 ± 0.107	0.652 ± 0.076
10	5	F, $\mathscr{A}_5$	3	0.627 ± 0.062	0.624 ± 0.081	0.629 ± 0.065

System FuzzyDataMiner

The *FuzzyDataMiner* program was created as a part of the implementation of the data analysis methods described in this book. Necessary information as well as the software itself can be obtained from [1]. The main purpose of this software is to present two applications of practical methods described in this book. The first application concerns the classification of test objects using a classifier constructed by the *k-NN* method in a situation where missing values appear in the test object. The second one is the classifier designed for microarray data, which is characterized by a very large number of attributes and a small number of objects.

The program has been implemented in Java. Therefore, for proper operation it is necessary to install the Java SE Development Kit 8 or at least Java SE Runtime Environment 8 from Oracle. To test the *FuzzyDataMiner* program, from the above mentioned web page, it is necessary to download the zip archive called *fuzzydataminer.zip*. After unpacking this file, the *FuzzyDataMiner* directory will appear on the disk containing the following files:

- *fuzzydataminer.jar* the runnable archive jar of the FuzzyDataMiner program,
- *biodeg.tab* an example of a data set to test a method with a parameter -S,
- *biodeg_config.txt* an exemplary configuration of 3 experiments to test the method with the parameter -S,
- *leukemia.tab* an exemplary data set to test the method with the parameter -M,
- *leukemia_config.txt* an exemplary configuration of 2 experiments to test the method with the parameter -S,
- *start_s.bat* a batch file to run the experiment with the method -S (this file can be run from the window of the Windows file manager without using the terminal),
- *start_m.bat* a batch file to run the experiment with the method -M this file can be run from the window of the Windows file manager without using the terminal).

U. Bentkowska, *Interval-Valued Methods in Classifications and Decisions*, Studies in Fuzziness and Soft Computing 378,
https://doi.org/10.1007/978-3-030-12927-9

In the *fuzzydataminer directory* there will also appear a *lib* subdirectory containing the WEKA API library version 3.8, which is used by the *FuzzyDataMiner* program. In addition, from the above mentioned page one may download two sets of the data sets used for experiments described in the book. The first collection named *SDataSets.zip* is a data set for experiments with the missing values feature. The other one, called *MDataSets.zip* is a microarray data set. More details concerning the *FuzzyDataMiner* program with the user manual are available at [1].

Reference

1. http://diagres.ur.edu.pl/~fuzzydataminer/

Index

U. Bentkowska, *Interval-Valued Methods in Classifications and Decisions*,
Studies in Fuzziness and Soft Computing 378,
https://doi.org/10.1007/978-3-030-12927-9

Printed in the United States
By Bookmasters